# THE RISE AND DECLINE
of U.S. Military Culture Programs,
2004–20

# THE RISE AND DECLINE

## of U.S. Military Culture Programs, 2004–20

*Edited by*
*Kerry B. Fosher, PhD,*
*and Lauren Mackenzie, PhD*

MARINE CORPS UNIVERSITY PRESS

Quantico, Virginia
2021

LIBRARY OF CONGRESS CATALOGING-IN-PUBLICATION DATA
Names: Fosher, Kerry B., editor. | Mackenzie, Lauren, 1976– editor. | Marine Corps University (U.S.). Press, issuing body.
Title: The rise and decline of U.S. military culture programs, 2004–20 / edited by Kerry B. Fosher and Lauren Mackenzie.
Description: Quantico, Virginia : Marine Corps University Press, 2021. | Includes bibliographical references and index. | Summary: "Though the priorities of senior military leaders inevitably change over time, the pressing need for American Service personnel to accommodate the human dimension for success in their ongoing military operations has not diminished. That capability now may be even more important than ever. Almost inevitably, the requirement will reach a critical stage in some future crisis. This book compiles the insights and findings of some of the most determined and resourceful scientists, scholars, and practitioners engaged in the military's culture programs to inculcate the new capabilities in the early twenty-first century. The authors do not gloss over failures and dead ends. Rather, their expectation is that by presenting the bad with the good, they can help future generations engaged in the same task avoid their pitfalls and build on their work. More importantly, the authors hope that their writing might reach those who are still engaged in building cultural capabilities and that they will find encouragement to continue this essential work"— Provided by publisher.
Identifiers: LCCN 2021021198 | ISBN 9781732003187 (paperback)
Subjects: LCSH: Cultural competence—Study and teaching—United States. | Intercultural communication—Study and teaching—United States. | Cross-cultural orientation—United States. | International relations and culture—United States. | Military education—Social aspects—United States. | United States—Armed Forces—Officials and employees—Training of.
Classification: LCC HM793 .R57 2021 | DDC 303.482—dc23 | SUDOC D 214.502:C 89 LC record available at https://lccn.loc.gov/2021021198

## DISCLAIMER

The views expressed in this publication are solely those of the authors. They do not necessarily reflect the opinions of the organizations for which they work, Marine Corps University, the U.S. Marine Corps, the U.S. Navy, the U.S. Army, the U.S. Air Force, or the U.S. government. The information contained in this book was accurate at the time of printing. Every effort has been made to secure copyright permission on excerpts and artworks reproduced in this volume. Please contact the editors to rectify inadvertent errors or omissions. In general, works of authorship from Marine Corps University Press (MCUP) are created by U.S. government employees as part of their official duties and are not eligible for copyright protection in the United States; however, some of MCUP's works available on a publicly accessible website may be subject to copyright or other intellectual property rights owned by non-Department of Defense (DOD) parties. Regardless of whether the works are marked with a copyright notice or other indication of non-DOD ownership or interests, any use of MCUP works may subject the user to legal liability, including liability to such non-DOD owners of intellectual property or other protectable legal interests.

The production of this book and other MCUP products are graciously supported by the Marine Corps University Foundation.

Published by
Marine Corps University Press
2044 Broadway Street
Quantico, VA 22134
1st Printing, 2021
ISBN: 978-1-7320031-8-7

THIS VOLUME IS FREELY AVAILABLE
AT WWW.USMCU.EDU/MCUPRESS

# CONTENTS

Foreword — vii
Preface — xiii
Acknowledgments — xvii

## INTRODUCTION — 3
*by Kerry B. Fosher, PhD, and Lauren Mackenzie, PhD*

## CHAPTER ONE — 19
Big Battles, Small Victories:
Personal Experience in Culture Wars, 2003–9
*by Ben Connable, PhD*

## CHAPTER TWO — 41
On Becoming "Wise in the Ways of Others":
Lessons Learned from Integrating Culture
into Professional Military Education Curriculum
*by Lauren Mackenzie, PhD*

## CHAPTER THREE — 56
From Aha Moments to Emerging Stories of the Good Old Days:
Reflections from Many Years in a Fascinating Field
*by Susan Steen, PhD*

| | |
|---|---|
| CHAPTER FOUR | 66 |
| Surfing the Sine Wave of Military Culture Education | |
| *by Angelle Khachadoorian, PhD* | |
| CHAPTER FIVE | 77 |
| The Company I Kept: | |
| Twenty Years at the Naval Postgraduate School | |
| *by Anna Simons, PhD* | |
| CHAPTER SIX | 104 |
| From Concept to Capability: | |
| Developing Cross-Cultural Competence | |
| through U.S. Air Force Education | |
| *by Brian R. Selmeski, PhD* | |
| CHAPTER SEVEN | 125 |
| Bridging the Social Science Research-to-Practice Gap | |
| *by Allison Abbe, PhD* | |
| CHAPTER EIGHT | 142 |
| A Few Things I Know about Culture Programs | |
| or Why Nothing Works | |
| *by Kerry B. Fosher, PhD* | |
| CHAPTER NINE | 162 |
| Alternative Perspectives: | |
| Launching and Running the Marine Corps' Culture Center | |
| *interviews with Jeffery Bearor and George Dallas* | |
| CONCLUSION | 203 |
| *by Kerry B. Fosher, PhD, and Lauren Mackenzie, PhD* | |
| APPENDIX | 209 |
| Common Culture Program Lines of Effort | |
| Select Acronyms and Terms | 213 |
| Bibliography | 219 |
| Index | 235 |
| About the Authors | 243 |

# FOREWORD

In the wake of America's trauma on 11 September 2001 (9/11), the nation embarked on an unprecedented series of struggles against perceived foes that had the capacity to bring terror to the nation's heartland. This soon led to military interventions in familiar places like Iraq and in very unfamiliar places like Afghanistan. Ultimately, American military involvement also would extend to such unlikely locations as Syria, Yemen, Somalia, and Mali.

The challenges of America's worldwide "full spectrum operations" were profound, though the nation's well-trained and well-equipped armed forces proved equal to the purely military tasks.[1] The intercultural challenges, conversely, were extraordinarily complex and problematic. At the same time, they were so critical to success that failure to accommodate the hu-

---

[1] For more on this concept of full spectrum operations, see *The Current Status of U.S. Ground Forces*, Hearing before the Subcommittee on Readiness and Management Support of the Committee on Armed Services, 111th Cong., 1st Sess. (22 April 2009); and John Morrissey, "Securitizing Instability: The US Military and Full Spectrum Operations," *Environment and Planning D: Society and Space* 33, no. 4 (2015): 609–25, https://doi.org/10.1068/d14033p.

man dimension obviated the value of any skillful military operation. And it was precisely in this area that American wealth and technology could not easily overcome the obstacles.

American military personnel soon found themselves in operational areas where the definitions of friend, foe, and noncombatant were inherently fluid, changeable with every event. Primary loyalties tended to emphasize patron-client networks, often kinship based. In addition to the natural barriers of language, local systems of logic produced interpretations of causality, obligation, motivation, and action that varied radically from the American experience. Short-term self-aggrandizement sometimes trumped any vision for a better society or functioning nation. Both friend and foe at times displayed a dismaying contempt for Western notions of *jus in bello*, or international humanitarian law. And these were just a few examples of the many challenges. By about 2003, as military operations in Afghanistan and Iraq ramped up, it had become glaringly evident to military personnel and defense civilian leaders that new capabilities were needed to deal with the human dimensions of the new military involvement.

Each of America's Armed Services recognized the need, and by 2005 each had initiated tentative efforts to address it. For example, the Air Force initiated activities that would eventually lead to its culture center and the Marine Corps was starting culture-related training and taking the initial steps in forming its center. Within another year, the responses started to take shape in the form of Service culture centers. A considerable amount of communication also had begun among the scholars, scientists, managers, and practitioners involved in these efforts, both informally and through a robust agenda of consultations and conferences. Many of the authors in this book met, talked by telephone, and corresponded frequently, sometimes daily, during the early years of growth of the culture centers. There were formal gatherings of the Department of Defense (DOD) "culture community," such as the annual conferences hosted by the Air Force, Army, and Marine Corps. There also were informal gatherings at scholarly conferences such as the Inter-University Seminar on Armed Forces and Society meetings. The frequent communication led to a tight-knit community that emphasized collaborative work.

There was little question that the evolving roles of the nation's soldiers, sailors, airmen, and Marines required a greater ability to communicate, collaborate, and influence in the culturally complex circumstances of the new environments.[2] It was much less clear at the outset how to define, build, and maintain those capabilities, or what could be expected within the constraints of time and resources. An initial, naïve (though pervasive) assumption in the DOD was that the problem simply was linguistics and that more emphasis on language learning would provide the necessary cultural understandings. Meanwhile, military leaders were pressing for immediate solutions, preferably involving capabilities that were deliverable as part of brief predeployment training sessions.[3]

Much clearer insights emerged slowly during the course of a decade from the research, experimentation, and innovation of small groups of brilliant scientists and scholars. They found some training measures could assist Service personnel in coping with the immediate cultural complexity of their assignments. But they also found that a true capability to

---

[2] Each of the respective Services formats references to their servicemembers in a variety of ways—Soldiers, Sailors, Airmen, and Marines. Throughout this work, and in an effort to remain consistent with Marine Corps University Press style conventions, we will be using Marine Corps standards for military terminology.

[3] John E. Kruse et al., *Building Language Skills and Cultural Competencies in the Military: DOD's Challenge in Today's Educational Environment* (Washington, DC: Subcommittee on Oversight and Investigations, U.S. House of Representatives Committee on Armed Services, 2008); and Paula Caligiuri et al., *Training, Developing, and Assessing Cross-Cultural Competence in Military Personnel*, Technical Report 1284 (Arlington, VA: U.S. Army Research Institute for the Behavioral and Social Sciences, Department of the Army, 2011).

communicate, collaborate, influence, and lead in culturally complex circumstances required a process of long-term education. The ability to suspend judgment, recognize patterns and cues, "see" reality through the filter of other world views, understand the differing expectations of leadership, and know how to incentivize behavior could only be acquired by deliberate, comprehensive, and progressive intellectual development.[4]

Military and civilian leaders in the Services could nod and smile in apparent agreement with the recommendations for education and professional transformation, but they never really wrapped their heads around what it would mean in terms of policy, planning, and resources. By the time the implications of these insights were fully understood, national priorities had changed, the interest of senior military leaders had waned, and many of the programs to develop these new capabilities were scaled back or abandoned. Ironically, this pattern had occurred at least twice in the previous half century. In the immediate aftermath of World War II and during the Vietnam conflict, the nation's military leaders had recognized much the same need and had initiated programs to address it, only to abandon them in the end.[5]

Most unfortunately, the institutional memory of these earlier efforts was vague indeed, consisting for the most part of obscure and anecdotal historical footnotes. The work behind them, including successes and failures, was largely unrecorded and irretrievable. This was an understandable but dam-

---

[4] Lauren Mackenzie, Eric Gauldin, and Erika Tarzi, *Cross-Cultural Competence in the Department of Defense: An Annotated Bibliography* (Quantico, VA: Center for Advanced Operational Culture Learning, 2018).

[5] For more on these programs during times of conflict, see Lauren Mackenzie and John W. Miller, "Intercultural Training in the United States Military," in *The International Encyclopedia of Intercultural Communication*, ed. Young Yun Kim and Kelly L. McKay-Semmler (Hoboken, NJ: John Wiley, 2017), https://doi.org/10.1002/9781118783665.ieicc0189.

aging oversight, particularly in view of the cyclical nature of the need and the almost inevitable likelihood that the requirement would become pressing again in time. In fact, the ability to work effectively in culturally complex environments ultimately may prove to be the most essential military capability of the twenty-first century.

There has been, nonetheless, a silver lining to the most recent retrenchments. Some of the infrastructure remains. Some scholars and scientists hired into the Service culture centers have migrated into professional military education (PME) institutions, where they continue to offer their unique expertise to generations of military students. One Service, the Air Force, deliberately embedded culture instruction into every level of military education in a way that may survive the vagaries of official educational emphasis. There is even some reason to hope that sufficient senior leadership vision remains to establish a Department of Defense center of excellence that could guide efforts to embed the development of cross-cultural competence into the whole fabric of Service education to preserve, nurture, and further develop what has been so painstakingly created.

Though the priorities of senior military leaders inevitably change over time, the pressing need for American Service personnel to accommodate the human dimension for success in their ongoing military operations has not diminished. That capability now may be even more important than ever. Almost inevitably, the requirement will reach a critical stage in some future crisis.

This book compiles the insights and findings of some of the most determined and resourceful scientists, scholars, and practitioners engaged in the programs to inculcate the new capabilities in the early twenty-first century. The authors do not gloss over failures and dead ends. Rather, their expectation is that by presenting the bad with the good, they can help future generations engaged in the same task avoid their pitfalls and build on their work. More importantly, the authors hope that their writing might reach those who are still engaged in building cultural capabilities and that they will find encouragement to continue this essential work.

*Colonel Daniel Henk, PhD*
*U.S. Army (Ret)*

# PREFACE

This project is intended to capture experiences and lessons learned during a remarkable time period, roughly 2004–20, during which the U.S. Department of Defense (DOD) experimented with a broad range of programs and initiatives related to culture. These programs were intended to improve military personnel's capacity to understand and operate effectively within culturally complex environments. While most culture-related efforts focused on conflicts in Iraq and Afghanistan, especially during the early years, they also helped address a capability gap in missions around the globe.

The cultural complexity these programs were intended to address arose from a number of different sources, including adversaries who did not think and behave as expected and whose motivations were opaque to U.S. personnel, local populations whose interpretations of events and behaviors did not align with expectations and whose good will and support were essential to mission success, and partners whose worldview was just different enough to cause misunderstandings. Even collaboration with other U.S. government agencies and nongovernmental organizations was sometimes culturally complex, as many military personnel came to understand when they tried to work with aid organizations. And woven through all these intercultural experiences was *change*. Military personnel encountered people who were adapting culturally to changes brought on by conflict, disaster, and development rather than adhering to the often outdated and inaccurate guidebooks and checklists circulating across the DOD. It was a messy problem set and, particularly in the early years, there was little consensus as to how to solve it.

Those of us working in culture programs recognized that the DOD's interest in culture was cyclical. There had been cycles of interest during World War II, the Vietnam War, and the Cold War era. Each time, interest waned after a few years and programs were dismantled. So, we knew that the cycle we were in would likely end, only to pick up gain in 5, 10, or 20 years.

Because the cycle of interest in culture repeats, as editors, we wanted to bring together a volume that not only described some of the activities that took place between 2004 and 2020, but also captured the ways in which people became and stayed involved, and reflections on their lessons learned. We made the decision to emphasize social science voices—although several chapter authors have military backgrounds—because when the DOD turns its attention toward culture, it typically realizes its pool of social scientists is relatively small and begins recruiting more, often from fields with little connection to the department. The friction that results from DOD civilians, military personnel, and new social scientists interacting with one another was an important piece of the context we wanted to capture.

Readers will note that most of the chapter authors have longstanding professional relationships (and sometimes friendships). While the selection of interconnected authors creates the risk of the book seeming self-referential, we hope readers will understand that the connections arose through the work at hand, including many years of collegial argument and consensus building. We also made the choice to focus the volume primarily on culture-related work in the education and training domain. This was in part to introduce balance to literatures that are dominated by books, articles, and reports about the U.S. Army's now defunct Human Terrain System and the DOD's technology-centered approaches to culture. However, the choice also reflects a pattern across DOD's culture efforts. The social scientists who stayed connected to the department's culture efforts over the years tended to gravitate toward education and training. This can be attributed partially to scientists finding more congenial working conditions in military colleges and universities, but it also reflects their choices about where to put their energies. Put bluntly, most of us felt like we were more likely to create long-lasting capability by influencing the thinking of and helping to develop military personnel rather than through systems and tools.

We deliberately chose authors who worked in different Services, different organizations, and different types of jobs. These diverse standpoints, combined with the various backgrounds of the authors, mean that each chapter brings a unique vantage point on similar challenges and diverse lessons learned from what went right and what went wrong. As mentioned in the introduction, we hope the experiences and lessons presented in this volume will be valuable to those currently working in culture programs, but also serve as an archive or message in a bottle for those who are involved in the next cycle of the DOD's interest in culture.

# ACKNOWLEDGMENTS

The editors would like to first thank our military and government civilian partners in creating and sustaining culture programs over the years. As social scientists, we often have the loudest voice in publications about DOD's culture efforts, but we are painfully aware that we only had one piece of the puzzle. We also would like to thank the authors who wrote these chapters in the midst of a global pandemic, many while juggling home schooling and teleworking on top of all the usual chaos of busy lives. We are especially grateful to George Dallas, Jeffery Bearor, and Daniel Henk for their willingness to reflect on their experiences leading DOD culture centers and for taking the time to engage with us in countless conversations over the years. For us, they have embodied the adage that *leadership happens in conversation*. Last, but definitely not least, we thank Angela Anderson and Marine Corps University Press for their receptivity to an unusual book project and their encouragement and support throughout the process of developing it.

# THE RISE AND DECLINE
of U.S. Military Culture Programs,
2004–20

# INTRODUCTION

*by Kerry B. Fosher, PhD,
and Lauren Mackenzie, PhD*

## The Rise

In the early and mid-2000s, the U.S. Department of Defense (DOD) was confronting challenges in Iraq and Afghanistan that military leaders understood to be cultural. Military personnel, trained and equipped for combat, found themselves in situations where they needed to understand and communicate with local people and partner forces from other countries. At all levels—individual military personnel, units, commands, and entire Services—people tried to find ways to improve understanding of what was commonly referred to as *culture*.

Almost two decades later, it is difficult to convey how intense and at times fractious these early efforts were. The main interest was in the urgent need to improve military personnel's ability to navigate the cultural complexity they encountered in Iraq and Afghanistan, to understand not just what people were saying but what it meant in a cultural context. Military leaders saw this capability as essential to being able to operate effectively, to anticipate how local populations would respond to U.S. and Coalition actions, and to attempts to influence local people's perception of U.S. forces and goals. However, the best way to build the capability was very much in debate.

Services competed for scarce subject matter experts to provide training or to deploy with units. Some people, savvy about the way the DOD works, took advantage of the situation to promote themselves and their pet programs.[1] Buzzwords like *cultural savvy* and *cultural astuteness* proliferated with each proponent ensuring worried leaders, sometimes with evangelical zeal, that their approach was the one true path to operational effectiveness. Existing organizations tried to make the case that their policies and programs related to regional knowledge and language were already providing the capability, despite the demand signal from the operating forces. The larger DOD science and technology bureaucracy did what it does and delivered a steady stream of projects to "solve the culture problem" with databases, models, simulations, and other technology-centric offerings.[2] Some parts of the Office of the Secretary of Defense pushed for centralized, Joint culture programs while other parts funded all kinds of niche culture initiatives. Meanwhile, social scientists and other experts inside and outside the department, some of whom have con-

---

[1] While we do not want to call out any specific individuals or programs, during this time period, there was a great deal of funding available for anything labeled "culture." Many people inside and outside the DOD with few, if any, qualifications, began labeling themselves as cultural experts. Others simply rebranded their existing efforts as somehow connected to culture in order to tap into the sudden rush of funds.

[2] See for example, Kerry B. Fosher, "Cautionary Tales from DoD's Pursuit of Cultural Expertise," in *Cultural Awareness in the Military: Developments and Implications for Future Humanitarian Cooperation*, ed. Robert Albro and Bill J. Ivey (Hampshire, UK: Palgrave Macmillan, 2014), 15–29, https://doi.org/10.1057/9781137409423; Human, Social, Culture, Behavior Modeling Program, *Human Social Culture Behavior Newsletter* 1, no. 1 (Spring 2009); and Jessica Glicken Turnley and Aaron Perls, *What Is a Computational Social Model Anyway?: A Discussion of Definitions, a Consideration of Challenges, and an Explication of Process* (Albuquerque, NM: Advanced Systems and Concepts Office, Defense Threat Reduction Agency, 2008).

tributed chapters to this volume, debated the right long-term approaches while trying to provide something to meet the immediate needs.

Over time, many of these efforts to train, educate, and advise military personnel coalesced into culture programs and centers. The most commonly known of these are the Service culture centers: the Army's Training and Doctrine Command (TRADOC) Culture Center at the U.S. Army Intelligence Center, Fort Huachuca, Arizona; the Air Force Culture and Language Center (AFCLC) at Maxwell Air Force Base, Alabama; the Navy's Center for Language, Regional Expertise and Culture in Pensacola, Florida; and the Marine Corps' Center for Advanced Operational Culture Learning (CAOCL) in Quantico, Virginia. There also were culture programs in the DOD research community, such as those run by the Army Research Institute, individual faculty who provided increased culture-related courses at military academies and universities, and new culture-focused training and analysis classes at places like the Air Force Special Operations School at Hurlburt Field in Florida and Marine Corps Intelligence Activity (MCIA) in Quantico. There was still a seemingly endless amount of funding for anything related to culture, so the proliferation of new initiatives, buzzwords, and approaches continued. Yet, the formal culture centers and programs did bring a sense of coherence and opportunities for coordination and collaboration.

During this same period, military organizations sought to recruit experts from academia. There was a particular focus on trying to hire anthropologists, but organizations also sought experts on the Middle East and Afghanistan and social scientists from other disciplines. Some of these academic fields had long memories of past encounters with the military that had been fraught with misunderstandings and ethical issues, and they were understandably wary of renewing the connections. There were debates about whether it was appropriate for scholars to provide support to the military.[3] These debates often were professional and well-informed, but sometimes devolved into personal attacks. It was a challenging time to be an academic working in a military context. The debates and heightened attention to ethics diminished an already small number of academics willing to work with the military, reducing the pool from which the military and defense contractors could hire. As discussed in some of the chapters that follow, limitations on the number of qualified experts available had both negative and positive outcomes. As one might expect in such circumstances, many unqualified individuals or outright charlatans were hired, which created problems for the military personnel they supported and for those trying to create con-

---

[3] The debates in anthropology are particularly well documented. See, for example, Robert Albro et al., eds., *Anthropologists in the Securityscape: Ethics, Practice, and Professional Identity* (Walnut Creek, CA: Left Coast Press, 2012); Robert Albro et al., *Final Report on the Army's Human Terrain System Proof of Concept Program* (Arlington, VA: American Anthropological Association [AAA] Commission on the Engagement of Anthropology with the U.S. Security and Intelligence Communities, 2009); Kerry Fosher, "Review Essay: Anthropologists in Arms: The Ethics of Military Anthropology," *Journal of Military Ethics* 9, no. 2 (2010): 177–81, https://doi.org/10.1080/15027570.2010.491357; Kerry Fosher, "Yes, Both, Absolutely: A Personal and Professional Commentary on Anthropological Engagement with Military and Intelligence Organizations," in *Anthropology and Global Counterinsurgency*, ed. John D. Kelly et al. (Chicago, IL: University of Chicago Press, 2010), 261–71; George R. Lucas, *Anthropologists in Arms: The Ethics of Military Anthropology* (Lanham, MD: Altamira Press, an imprint of Rowman & Littlefield, 2009); Laura A. McNamara and Robert A. Rubinstein, eds., *Dangerous Liaisons: Anthropologists and the National Security State*, School for Advanced Research Advanced Seminar Series (Santa Fe, NM: School for Advanced Research Press, 2011); James Peacock et al., *Final Report, November 4, 2007* (Arlington, VA: AAA Commission on the Engagement of Anthropology with the U.S. Security and Intelligence Communities, 2007); David H. Price, *Weaponizing Anthropology: Social Science in the Service of the Militarized State* (Oakland, CA: AK Press, 2011); and Robert A. Rubinstein, "Master Narratives, Retrospective Attribution, and Ritual Pollution in Anthropology's Engagements with the Military," in *Practicing Military Anthropology: Beyond Expectations and Traditional Boundaries*, ed. Robert A. Rubinstein, Kerry B. Fosher, and Clementine K. Fujimura (Sterling, VA: Kumarian Press, 2012), 119–33.

ceptually and factually sound programs. On the positive side, the small number of social scientists across DOD culture efforts also led to significant collaboration, which in turn made it possible to coordinate to achieve greater effects on policy and programs than normally would have been possible.[4]

It was in this messy and intense context that many of the authors, some of whom already knew one another, began working together.

## Collaboration and Connection

Each of the contributors to this volume entered into work with DOD culture programs on a different trajectory and are connected in various ways. Kerry Fosher (chapter 8) and Brian R. Selmeski (chapter 6) met in graduate school at Syracuse University in New York State and began working together more frequently around 2006. Both knew Anna Simons and had been influenced in one way or another by her research and writing over the years (chapter 5). Fosher also was using Ben Connable's (chapter 1) article, "Marines Are from Mars, Iraqis Are from Venus," in an introductory anthropology class before she knew him.[5] Selmeski then ran across Connable at a conference at Brown University in Providence, Rhode Island, and put them in touch by phone. The three met formally at the Inter-University Seminar on Armed Forces and Society (IUS) that same year.

In late 2006, Selmeski and Fosher began to help the Air Force with their culture efforts in an informal capacity. In January 2007, Fosher started working for Air University at Maxwell Air Force Base. She and Selmeski traveled a great deal for

---

[4] For example, social scientists across the Services worked together to shape the way the Defense Language Office in Washington, DC, would approach culture. They also collaborated on inputs to new or revised policy and doctrine and on developing realistic approaches to assessing the effectiveness of education and training curricula.

[5] Maj Ben Connable, "Marines Are from Mars, Iraqis Are from Venus," *Small Wars Journal*, 30 May 2004.

that work, and it was around then that they met Allison Abbe (chapter 7) and Jeffery Bearor (chapter 9). Fosher met George Dallas (chapter 9) in 2008, when he became the director of CAOCL and she was working for the Marine Corps Intelligence Activity.

Lauren Mackenzie (chapter 2) began working as a faculty member in Brian Selmeski's Cross-Cultural Competence (3C) Department at AFCLC in 2009 and met Fosher that same year at the IUS conference. Angelle Khachadoorian (chapter 5) joined the 3C department in 2011 after having served as a faculty member at the U.S. Air Force Academy in Colorado Springs, and Susan Steen (chapter 4) replaced Mackenzie as a 3C Department faculty member at the AFCLC in 2014 after working for many years in international education. Mackenzie began working with Fosher at CAOCL in 2015, where they served as leads in teaching and research, respectively, until CAOCL closed its doors in 2020.

The authors' different trajectories and connections affected not only their specific jobs and work, but their sense of the overall effort, what mattered, what was possible, and the size and shape of the roles they could play. The work itself has changed over the years. In some cases, these changes resulted from taking a new job, in some cases because the nature of the work within a job changed, and in others because their thinking evolved as they learned more about the challenges and opportunities, the organizations in which they work, and the people with whom they work. Throughout, the authors have found ways to collaborate, argue congenially, and generally support one another's efforts to take advantage of this time when it was possible to make significant changes of benefit to military personnel and those with whom they interact around the world.

## The Culture in Culture Programs

Most of the programs in which the authors worked were in some way associated with the word *culture*. Various understandings of the term drove the development of programs and initiatives and guided their efforts. However, those early understandings often were at odds with contemporary science related to culture (as well as with the experiences of many military personnel), portraying it as a static set of traits, behaviors, or social structures that could be clearly described, a sort of system of predictably interacting parts. This view of culture worked for standardizing programs, writing policies, and creating slide decks, but it was not particularly useful for preparing military personnel for the fluid, changing cultural patterns they would actually encounter. Contemporary science rarely uses the term culture in an explanatory way. Rather, it is used as an umbrella concept to refer to the more complex relationship between individual human agency and the patterns of meaning and behavior that used to be labeled as culture.[6] For the practical purposes of the military, what matters is not a static description of "a culture" at a given point in time, but rather the way people enact, transform, and maintain those patterns. Inconvenient though it may be, that complexity is what military personnel encounter, not a group of people behaving in lock step with a set of rules. As the anthropologist, Tim Ingold, wrote, it is "more realistic to say that . . . people *live culturally*, rather than that they *live in cultures.*"[7]

As these things tend to go in DOD, definitions of culture proliferated in the early years. The flood of definitions occasionally was punctuated by pressure for individuals and organizations to settle on one definition that could be used in policy or doctrine. While we understood military organizations' interest in settling on one definition, we also knew that a definition of a term like culture would constrain more than guide, providing little help in building programs or curricula.

---

[6] See, for example, Tim Ingold, "Introduction to Culture," in *Companion Encyclopedia of Anthropology*, ed. Tim Ingold (New York: Routledge, 1994), 330, https://doi.org/10.4324/9780203036327.

[7] Ingold, "Introduction to Culture," 330.

Many of us echoed Brian Selmeski's early concern that the search was "a fool's errand certain to take a long time while producing a result of both questionable validity and utility."[8] It was more useful to focus on the specific aspects of culture that were salient for a particular mission and to teach the generalizable concepts and skills that could be used anywhere. No short definition could capture that complexity in a useful way.

Supposedly authoritative definitions did sometimes appear in policy and doctrine, although they rarely leveraged the expertise of the social scientists and others who had been brought in to guide culture programs. Other guiding documents mercifully addressed culture without defining it.[9] Fortunately, where definitions or descriptions existed, they typically left enough wiggle room to map our efforts onto them rather than being overly constraining. Within individual programs, we were usually successful in focusing attention on conceptual frameworks and approaches, which could guide practical efforts, rather than definitions, which had little utility.

Perhaps the most useful thing to come out of all the churn about definitions was confidence that they did not matter. Each of the authors in this volume likely has a somewhat different approach to culture based on disciplinary background and professional experience. Nonetheless, we found enough common ground to collaborate and stay focused on helping military personnel learn the concepts, skills, and culture-specific details they needed.

## The Decline

Most of the authors in this book knew early in their work that

---

[8] Brian R. Selmeski, *Military Cross-Cultural Competence: Core Concepts and Individual Development*, Armed Forces, and Society Occasional Paper Series No. 1 (Kingston, ON: Centre for Security, Armed Forces and Society, Royal Military College of Canada, 2007), 3.

[9] See, for example, *Chairman of the Joint Chiefs of Staff (CJCS) Instruction 3126.01A, Language, Regional Expertise, and Culture (LREC) Capability Identification, Planning, and Sourcing* (Washington, DC: CJCS, 31 January 2013).

the military's interest in culture and related social science was ephemeral. There were episodes of interest during World War II and the Vietnam era that led to the development of centers, programs, and classes.[10] Ultimately, these capabilities were dismantled as the military decided it was never going to do *that* again, with *that* being whatever type of warfare military leaders most closely associated with the need to learn about culture. Knowledge of past efforts, whether through scholarly accounts or bureaucratic records, was critical to our ability to avoid repeating past mistakes. Not everyone was interested in learning from this history. Especially in the early years when budgets were flush and the landscape was wide open, many people were excited about being the first or unique, or simply wanted to believe that conditions were so different now that no lessons from the past could be relevant.

However, some of us wanted to understand how and why we came to inherit such an open playing field rather than one occupied by robust culture programs that had been running for decades. Efforts to integrate social science in the Vietnam era have been extensively documented by scholars including the book *The Best-Laid Schemes: A Tale of Social Research and Bureaucracy* by Seymour J. Deitchman.[11] *The Best-Laid Schemes* was an important resource for many of us in understanding what had gone wrong before and envisioning different paths. Originally published in 1976, it captures the author's experiences with the military's efforts to integrate social science research and cultural knowledge during the 1960s. Perhaps more importantly for our purposes, it chronicled the gradual decline and dissolution of these efforts. The book was not a clear roadmap for us, and each reader found points of dis-

---

[10] Allison Abbe and Melissa Gouge, "Cultural Training for Military Personnel: Re-visiting the Vietnam Era," *Military Review* 92, no. 4 (July/August 2012): 9–17.

[11] Seymour J. Deitchman, *The Best-Laid Schemes: A Tale of Social Research and Bureaucracy*, 2d ed. (Quantico, VA: Marine Corps University Press, 2014; original printing by MIT Press, 1976).

agreement with the author. Regardless, it is a treasure trove of cautionary tales about the difficulty of effectively merging what social science has to offer with what DOD needs (which is not always the same as what it asks for).

The book also was out of print and copies were scarce. Seeing it as a resource for scholars and practitioners alike, Kerry Fosher worked with Deitchman and Marine Corps University Press to have the book reprinted in 2014. In her foreword to the new edition, she wrote, "Despite some successes, DOD is indeed on the verge of making many of the same mistakes Sy [Deitchman] documented, with consequences that will be borne by future junior military personnel and the people they encounter. There is still time to make course corrections. It is not quite time for one of us to write a sequel to *The Best-Laid Schemes*."[12] Six years later, it is time.

The authors started work on this volume as the entire context of DOD culture programs was changing. In fact, we started the work because of those shifts. Across the DOD, institutional interest in culture is waning; although, in some cases, the interest of individual military personnel is on the rise. With fading institutional interest come declining budgets, cuts in personnel, and sometimes closure.[13] The Service culture centers have been significantly affected by budget and personnel cuts, changes in focus away from culture and toward regional studies or language, and a general decline in senior leader attention needed to protect relatively new or unusual

---

[12] Kerry Fosher, "Foreword," in *The Best-Laid Schemes*, 4.
[13] For example, the Marine Corps' culture center has been closed and the Army and Navy centers have been significantly reduced in size and scope. Over the years, many other culture-focused programs and initiatives have been shuttered or repurposed, such as the Human Social Culture Behavior Modeling Program, the Army's Human Terrain System and Culture and Foreign Language Advisor Program, the Army's University of Foreign Military and Cultural Studies, the intelligence community's Socio-Cultural Dynamics Working Group, the culture research thrust area in the Army Research Institute, and Marine Corps Intelligence Activity's Cultural Intelligence Program to name only a few.

capabilities. This can be ascribed in part to shifting priorities within the DOD, which at the current moment is focused on the concepts of great power competition and information operations.[14] It is difficult to imagine how either of these concepts can be put to military use without personnel educated and trained to understand culture, but that logic has proven unpersuasive. Even without these new concepts, there is a healthy dose of the same "we're never going to do *that* again" attitude seen in past cycles. This time, the *that* was anything related to counterinsurgency, small wars, stability operations, and the like. Despite great efforts from many civilian and military personnel, we had largely failed to get across the basic point that the ability to understand and operate effectively in almost all missions requires an understanding of the cultural patterns of partners, local populations, and adversaries.

## Intent and Organization

The preceding paragraphs may seem dire, but the authors believe there are two reasons for hope. First, some culture programs are fairly well institutionalized and may continue long into the future, and other capabilities have been hidden away under names that do not mention culture for safekeeping. For those involved with current efforts, we offer this book as a collection of lessons learned from thoughtful colleagues with a broad range of experiences building and sustaining culture-related capabilities.

Second, even if every program was shut down, history sug-

---

[14] See, for example, David Vergun, "Great Power Competition Can Involve Conflict Below Threshold of War," Department of Defense, 2 October 2020; *Renewed Great Power Competition: Implications for Defense—Issues for Congress* (Washington, DC: Congressional Research Service, 2021); *Department of Defense Strategy for Operations in the Information Environment* (Washington, DC: Department of Defense, 2016); *Defense Primer: Information Operations* (Washington, DC: Congressional Research Service, 2020); and Michael Schwille et al., *Improving Intelligence Support for Operations in the Information Environment* (Santa Monica, CA: Rand, 2020), https://doi.org/10.7249/RB10134.

gests that they will be recreated in 5, 10, or 15 years and next time there will be a difference. In the beginning of the current cycle, while there were plenty of books and articles about the past written by civilian academics, it was difficult to find accounts and materials from inside military organizations. That was one of the reasons *The Best-Laid Schemes* was so valuable. Technological advances and deliberate efforts to archive materials should make it much easier for the next round of cultural capability developers to build on, rather than reinvent, past efforts. For those future readers, we hope this volume serves as a sort of message in a bottle and that some of its contents may save steps and stumbles in the construction of cultural capabilities.

The book's chapters reflect the diversity of backgrounds and experiences of the authors and are organized to take advantage of overlaps and differences. They capture decades of experience with different aspects of conceptualizing, building, refining, promoting, and defending military cultural capabilities. Broadly speaking, the chapters are organized along programmatic, teaching, and personal narrative themes. Because these chapters represent the singular, personal experiences of the authors, their thoughts are presented in first person to engage the reader and drive home the lessons in a way that a straight scholarly monograph might miss given the density of its language. Two chapters provide the perspective of former military personnel who have experienced both the user and provider sides of culture programs. Three chapters focus predominantly on issues related to teaching and learning. Two chapters emphasize programmatic issues and two blend lessons learned about teaching and programmatic concerns.

In chapter 1, Ben Connable, a retired Marine major and now PhD in war studies, begins with his experiences as a U.S. Marine in Iraq in 2003, tracing his growing sense of the urgency of improving Marines' understanding of culture. This chapter shows his evolution of thought through his deployments and tours in the supporting establishment, including his involvement in establishing the Marine Corps' culture center. He shares lessons from the classroom and from trying to sell the capability to an often-reluctant Service. He also describes the struggle between training and education programs and others, such as the Army's Human Terrain System, that sought to provide cultural *knowledge* without transforming the *thinking* of military personnel.

Chapters 2, 3, and 4 focus primarily on teaching and learning. In chapter 2, Lauren Mackenzie, a communication scholar, describes how she created a role for herself as a culture educator for military students. She offers examples and knowledge gleaned from experiences across different branches of Service, learning levels, and teaching modalities. She also provides recommendations for those who find themselves charged with the task of integrating culture content into professional military education (PME) curricula. Susan Steen, a communication scholar, uses chapter 3 to describe how she incorporates insights from the field of intercultural communication into her teaching at Air University. She provides three imperatives for readers that have informed the way she approaches culture education. In chapter 4, Angelle Khachadoorian, a cultural anthropologist, highlights the challenges associated with balancing her scholarly identity as an anthropologist with the (sometimes competing) expectations often required of military faculty members. She describes the ways in which she maintained boundaries to manage that tension and provides her lessons learned about connecting to military students in culture classes.

The authors of chapters 5 and 6 form a transition from the classroom to programmatic issues. Anna Simons, a social anthropologist, traces the very early roots of her interest in the military's cultural capability and her eventual decision to leave traditional academia to teach at the Naval Postgraduate School in chapter 5. Drawing on more than 20 years of experience with military education and research, she conveys insights on the transition to teaching in a military organization, facilitating culture-related learning for military students, the importance of research in influencing policy and programs, and the challenges of choosing the right opportunities. In chapter 6, Brian Selmeski, a cultural anthropologist, outlines his path to becoming involved in DOD culture efforts and his work in building the Air Force's culture center, including linking the concept of cross-cultural competence to the renewal of Air University's accreditation to award degrees. He also captures many early lessons learned from the policy arena and challenges related to building a sustainable program.

Chapters 7 and 8 are focused squarely on programmatic issues. In chapter 7, Allison Abbe, a social and personality psychologist, reflects on a career that produced some of the critical research that shaped culture programs during the last two decades. She outlines her move from traditional academia to a career focused on Army research, much of which was related to culture. She describes her efforts in the realms of training and education, science and technology, and bridging divides between academia and the military. She also highlights the importance of knowledge of the doctrine and strategic guidance processes for getting research adopted and conceptual approaches integrated. In chapter 8, Kerry Fosher, a cultural anthropologist, describes how her gradual shift to working with military organizations and involvement in academic debates about the ethics of working with the military shaped her approach to culture programs across her work with the Air Force and Marine Corps. She discusses the influence of bureaucratic gravity and discourses on efforts to reshape military organizations to accommodate culture-related concepts and lessons learned about integrating experts and expertise. She also emphasizes the benefits of collegial relationships and coordination in creating change.

The final chapter in the book, chapter 9, comprises two interviews with retired Marine colonels whose subsequent civilian careers involved running the Marine Corps' culture center, CAOCL. The first interview is with Jeffery Bearor, who was involved in the Service's efforts to build cultural capability while he was still in uniform and who transitioned to become the first civilian director of CAOCL in 2006. The second interview is with George Dallas, who ran the center from 2008 until its closure in 2020. These interviews provide overlapping but distinct perspectives on the challenges, successes, and failures of a major organization in DOD's overall culture effort. They also provide key lessons learned and suggestions for those who may be tasked with building or rebuilding culture-related capabilities in the future.

## CHAPTER ONE

# Big Battles, Small Victories
## Personal Experience in Culture Wars, 2003–9

### by Ben Connable, PhD

## Introduction

This chapter describes my personal experiences as a U.S. Marine foreign area and intelligence officer through three tours in Iraq, as one of the leaders of the Marine Corps' cultural training and intelligence efforts in the post-9/11 decade, and as a combatant in the bureaucratic wars over Department of Defense (DOD) culture dollars. In the mid- and late-2000s, I became a vocal opponent of the Human Terrain System (HTS) program run by the U.S. Army's Training and Doctrine Command.[1] My failed advocacy against HTS and on behalf of organic, force-wide cultural learning continues to shape the way I think, teach, and write about culture in 2020.

## Into the Fire

In early 2003, I was a long-in-the-tooth captain serving on then-brigadier general John F. Kelly's Marine task force during the invasion of Iraq. Kelly was tasked to push north from

---

[1] Jacob Kipp et al., "The Human Terrain System: A CORDS for the 21st Century," *Military Review* (September–October 2006): 8–15.

Baghdad to seize the city of Tikrit.[2] After brief fighting on the way into the city, the Marines attempted to stabilize the area. We had completely eliminated the Iraqi government. There was no army. There were no police, no doctors, no sanitation workers—only chaos. Thousands of Iraqis were looting government offices and military warehouses. Rifles, rocket launchers, and machine guns lay unattended in warehouses or dropped on the ground by fleeing Iraqi soldiers. Tens of thousands of civilians were trying to flood back into Tikrit across the bombed bridge over the Tigris River. Mixed in among this massive throng of agitated Iraqis were some men intent on exacerbating the chaos. Cultural obstacles and opportunities were unfolding in real time with life and death implications.

As a newly minted Middle East foreign area officer fresh out of a master's program and Arabic training, this was my first big chance to apply cultural knowledge to a real-world conflict. What followed was a roller-coaster ride of exhilarating, no-safety-net adaptation, along with many personal mistakes. A few weeks in Tikrit in 2003 left me with an appreciation for what goodwill, learned intuition, and perseverance can do in the absence of training and experience. It also gave me exposure to the mind-boggling close-mindedness and linear thinking of some of my fellow officers. Many people—Americans, Coalition partners, and Iraqis—suffered and died due to the failure of cultural training, education, and mindset in Tikrit in early 2003.

## From Monterey to Iraq

My road to Tikrit started in California. I was in the Arabic language course at the Defense Language Institute in Monterey on 11 September 2001. Soon after graduation, my cohort of foreign area officers would be thrust right into the mix in

---

[2] LtCol Michael S. Groen et al., *With the 1st Marine Division in Iraq, 2003: No Greater Friend, No Worse Enemy* (Quantico, VA: Marine Corps History Division, 2006), 369, hereafter *No Greater Friend, No Worse Enemy*.

Afghanistan and Iraq as the vanguard of Middle East cultural expertise for the Department of Defense.

We had arrived at the pinnacle of the military cultural training and education pyramid. We had earned advanced degrees in regional culture and history, spent 16 months in Arabic language training, and lived in places like Cairo, Egypt, or Amman, Jordan, for a year of regional immersion. Even with this education and experience, we were Middle East neophytes. Soon, we would be trying to train, educate, and advise people with even less cultural understanding as they fought to stabilize countries in the throes of insurgencies.

I traveled from Monterey to Cairo, where I spent a year improving my language skills and traveling around the Middle East. When I returned to the Marine Corps Headquarters in early 2003, I volunteered to deploy straight out to the 1st Marine Division. Then-major general James N. Mattis's Marines were sitting in the Kuwaiti desert preparing to invade Iraq. I knew I had to be there, but I did not know quite what I would do when I arrived. Just more than a week after I showed up at the forward command post, we launched our attack.

## Invading Iraq in 2003

I spent the next weeks playing interpreter, ersatz human intelligence officer, and learn-on-the-fly cultural advisor. I spoke with hundreds of Iraqis along the way north: Shi'a laborers who showed me the scars on their bodies from Saddam Hussein's torture chambers, terrified military prisoners who had shucked their uniforms in their haste to flee our attack, elderly people who thought that we were the Russians (or maybe even the Nazis), sick and frightened children, a few captured Iraqi generals, and, as we moved farther north, tribal elders.

My first meeting with an Iraqi tribal elder while on patrol in ar-Rashidiya, Iraq, would start a nearly two-decade-long relationship with mostly Sunni Arab tribal figures. I can still clearly recall the first meeting, standing on the street next to an open sewage trench. As the elder told me all of the things his neighborhood needed—water, sanitation, food, security—I scrambled to understand his dialect. I was fascinated and intrigued, sensing a whole world in this one part of Baghdad, Iraq, that I did not comprehend. In the end, I did little more than anyone else would have done. I recorded his concerns, reported them, and moved on.

In retrospect, I could have done so much more. I could have asked more thoughtful and probing questions. I could have personalized the discussion and stretched it out, and perhaps I could have moved it to a more formal setting. I could have helped Marine leaders get a handle on the situation in the streets of ar-Rashidiya. I did none of these things, primarily because though I had plenty of cultural education and language training, I had no meaningful cultural training and the wrong mindset for the task.

In early 2003, I could tell you all about the history of the Middle East. I could do a barely passable job of interpreting a conversation in Arabic. But nothing in my years of foreign area officer instruction suggested the best approach to what would later be called key leader engagements.[3] None of my education or immersion experience prepared me to understand this tribal elder as a person. In the years that followed, I became more and more convinced that history lessons and language skills mattered a lot less than a robust, general education in human behavior.

## Up to Tikrit, Iraq

A week after that discussion in Baghdad, we were in Tikrit. I drove around the city with a small human intelligence team collecting information that might help us capture Saddam Hussein. We started to pick up worrying threads of information on scheming and weapons hoarding by former regime leaders. In those first days, my primary duty in Tikrit was to

---

[3] For more on key leader engagement, see *Public Affairs*, Joint Publication 3-61 (Washington, DC; Joint Chiefs of Staff, 2016).

help Brigadier General Kelly get a handle on the population and to help set up a temporary governing body. I drove with him around Salah al-Din Province, all the while learning from his perceptive engagements with Iraqis.

Here was a light armor infantry officer with no cultural training doing everything I imagined a cultural expert should do. He waded into crowds with no body armor or weapon, putting his life at risk to extend a hand of friendship. He empowered me to meet with senior tribal elders on his behalf. He set aside valuable time to build relationships with local leaders, and he listened carefully when they spoke.

On the advice of some tribal elders, Kelly had his Marines put away their body armor and sling their weapons to demonstrate goodwill. We slept on the hard marble of the front porch of Hussein's palace rather than on the soft beds inside to prove to the Iraqis that we were not the next dictators of Iraq. Violence dropped precipitously, though its threat was omnipresent.

I helped another Marine infantry officer, Lieutenant Colonel Duffy W. White, set up a local police force. We held a recruiting drive. One Iraqi man jumped the security fence in his enthusiasm to sign up. Marines detained him and brought him to face White. Instead of issuing a knee-jerk punishment, Lieutenant Colonel White asked, "If you were a police officer and you caught someone jumping the security fence, what would you do?" We liked his answer and hired him.

I watched Kelly, White, and other officers who had no cultural training feel their way through these engagements with empathy and humility. I decided to emulate their approach. I took personal risks operating under the assumption that Iraqis would not hurt me if I was clearly trying to help them. I probably survived mostly by shocking Iraqis into inaction. In retrospect, getting into a car alone with four armed Iraqis for a drive into the countryside was probably ill-advised. But I survived, and by showing trust, I was able to gain trust.

Working from my observations of fellow Marines, I forged

relationships with a few tribal elders. That led me to greater opportunities to allay some of the violent intent bubbling under the surface in northern Iraq. I set up a series of meetings at a farm near Bayji, Iraq, some of which Kelly attended. I gave a speech in Arabic to about 200 notables, Kurds and Arabs, in which I tried to promise a new constitution for Iraq. I wound up mixing my Arabic words and promising a new notebook for everyone. It did not matter. What did matter is that I was reaching out to them and listening in return.

## Success and Failures in Bayji, Iraq

Throughout those first days in Tikrit, we never got near the city of Bayji. Our maps showed an Iraqi *Special Republican Guard Commando* unit sitting right in the city center. We could not know if they were still there, but we knew we had to secure the city. Kelly prepared to send an entire battalion of light armored vehicles in a tactical movement to probe Bayji to see if the commandos were still there. At the last moment, one of the tribal elders we had spent time with—the owner of the farm—told us that the Iraqi Army was gone. The people of Bayji had elected a new mayor and police chief. They were waiting to greet us with a welcoming committee and a party.

Kelly took great risk in accepting this information. Instead of executing a large tactical movement, we drove up the main road with about 100 Marines for security. Indeed, as the man promised, there was the mayor and chief of police with tea and cake and a big banner welcoming the Marines.[4] Along for the ride was a sharp U.S. Army captain who was our liaison officer from the division that was rapidly approaching from the south to relieve the Marines. Together, we cracked open a pen and used the ink to form a seal on a hastily written treaty of friendship between the pseudo governments of Bayji and Tikrit.

It was naïve to conclude that we had solved any problems

---

[4] Groen, *No Greater Friend, No Worse Enemy*, 369.

that day. But it felt like we were moving in the right direction. The next few days shattered that illusion.

For some inexplicable reason, the leaders of the U.S. Army division believed they were on a mission to rescue Kelly's Marines in Tikrit.[5] The morning they arrived at the palace, I was sitting in a folding chair, enjoying a cup of coffee, and looking out over the Tigris River. Soldiers in combat gear poured out of infantry fighting vehicles and, rifles up, cleared our bivouac area of any lurking threats. I am not sure who looked sillier, me in my skivvies in a lounge chair or the guys in heavy body armor assaulting our campsite. It was a troubling start to our transition.

The next day, I tried to introduce some of the officers of the division to the local tribal leaders. A senior officer told me that was not necessary, as they were not going to speak to them. Another senior officer told us that they "were going to show these [expletive for Iraqis] a thing or two."[6] A massive M1 Abrams main battle tank was positioned in the middle of one of the main thoroughfares, its 120mm barrel depressed to aim into oncoming civilian vehicles. Barbed wire went up all over the city. Gunfire sparked up as the division's soldiers shot at looters, sometimes justifiably and accurately, and other times less so on both counts.[7]

Most disturbing was the news we got from one of our liaison officers the night before we were scheduled to depart: the

---

[5] This information was relayed to the author by one of the unit's liaison officers in April 2004. Reference to this situation is made in Groen, *No Greater Friend, No Worse Enemy*, 396.

[6] This statement was made to a group of Task Force Tripoli officers by a senior leader in the Army division. This quote was written in full, with the original wording, in the first draft of the official history of the 1st Marine Division in Iraq in 2003. It was removed by editors prior to publication. The author reviewed the first draft of the report in which the quote was published. LtCol Michael S. Goren, "With the 1st Marine Division in Iraq, 2003: No Greater Friend, No Worse Enemy" (unpublished draft, Marine Corps History Division Occasional Paper, 2006).

[7] Groen, *No Greater Friend, No Worse Enemy*, 369–70.

U.S. Army division was planning to assault Bayji. They took their electronic map at face value and assumed the Iraqi commandos were still there. Kelly told them that was not the case and described our friendly tea party with the mayor and chief of police. General Kelly's initial comments fell on deaf ears, but we did manage to avert disaster that night. Days after we withdrew down to southern Iraq, I watched televised images of the division's soldiers kicking down doors in Bayji as they cleared the city of some nebulous threat.[8]

Within weeks, Bayji and Tikrit became hotbeds of anti-American violence.[9] Increasingly through the end of 2003 and into 2004, gunfire and roadside bombs greeted the soldiers as they drove through the city in armored vehicles. Detention sweeps picked up hundreds of young Iraqi men, some who were acting suspiciously and some who just happened to be nearby. I would meet many of these now-radicalized men during my visits to Abu Ghraib prison in 2004.[10]

Of course, I cannot prove that the hostile, close-minded mentality of some—certainly not all—of the leaders of that division led to the chaos that followed in Tikrit, Bayji, and the surrounding area. They did correct a grievous error we made. In our haste to normalize Tikrit, we did not secure all of the abandoned Iraqi weapons warehouses. Overabundance of culturally attuned goodwill armed some of the men who would go on to become insurgents. But it is an open question as to whether those men would have used those weapons if they

---

[8] Rory McCarthy and Ewan MacAskill, "US Steps Up Aggression in Tikrit," *Guardian*, 17 November 2003; Thomas E. Ricks, "'It Looked Weird and Felt Wrong'," *Washington Post*, 24 July 2006; Antonius C. G. M. Robben, "Chaos, Mimesis and Dehumanisation in Iraq: American Counterinsurgency in the Global War on Terror," *Social Anthropology* 18, no. 2 (2010): 138–54, https://doi.org/https://doi.org/10.1111/j.1469-8676.2010.00102.x; and Ann Scott Tyson, "Iraq's Restive 'Sunni Triangle'," *Christian Science Monitor*, 24 September 2003.

[9] McCarthy and MacAskill, "US Steps Up Aggression in Tikrit"; and Scott Tyson, "Iraq's Restive 'Sunni Triangle'."

[10] Seymour M. Hersh, "Torture at Abu Ghraib," *New Yorker*, 30 April 2004.

had not been so purposefully and aggressively alienated by the unit that replaced Kelly's Marines in Tikrit.

## A Grassroots Movement Emerges in the United States

In late May 2003, I returned to my desk in Washington, DC. Driven by my mixed experiences in Iraq—including observing several outright cultural failures that led to death—I tried to find ways to help improve military cultural training and education. I started writing and reaching out, looking for like-minded colleagues.

Thankfully, I was not alone. A small group of Army and Marine officers and social scientists found each other through workshops and conferences during the next year. Each was driven by personal experiences and by the reports of errant behavior by U.S. servicemembers in Iraq and Afghanistan. We all generally agreed on a few points in our face-to-face meetings and in a robust online forum.[11]

First, there was effectively no cultural training in the military in the early 2000s. What had been built during the Vietnam War was long gone. Second, lack of training and cultural knowledge was undermining the campaigns in Iraq and Afghanistan. Third, the social science community was going to have to find a way around its long-standing wariness of the military in order to help, in our words, "reduce the necessity for the use of violence" in war.[12]

---

[11] These bulletin board-style emails and posts constitute one of the richest narratives of the early efforts to develop cultural training and education capabilities in the U.S. military. Unfortunately, these posts were made under the assumption that they would remain private. They cannot be cited here and may never see the light of day.

[12] Ben Connable, "All Our Eggs in a Broken Basket: How the Human Terrain System Is Undermining Sustained Cultural Competence," *Military Review* (March–April 2009).

## Back to Iraq in 2004: ar-Ramadi

Before we could make much progress on the home front, I deployed back to Iraq as Mattis's cultural advisor. I again joined the Marines in Kuwait and set up ad hoc training in basic cultural awareness. It was clear from my first days in the desert with the Marines that there had been few improvements in cultural training or education since the invasion in early 2003.

Mattis and his staff had prepared to run a culture-conscious operation based on 80 stability operations rules. Marines were told to treat the population with respect. One of the points read, "Dignity and distance is the best way to treat Iraqi women."[13] These were all important but relatively basic points. Ideally, these would have been a set of starting points for a more nuanced plan.

From January to August 2004, I served as Mattis's cultural advisor, working out of the intelligence office in ar-Ramadi. I worked directly with the provincial governor and set up almost daily meetings with government officials and tribal elders across the region.

We had a rude introduction to al-Anbar Province in January 2004. It was clear when we arrived that the soldiers we were replacing had mentally burned out and lost their patience with the population. Some of their behavior was indicative of both a failure of cultural training and education. It also revealed an institution-wide failure to appreciate the importance of a thoughtful, patient mindset in complex operations like the Iraqi counterinsurgency.

On the drive in from Kuwait to ar-Ramadi, the Army machine gunner in our escort vehicle rapidly shifted the muzzle of their gun back and forth, aiming down into the cars of terrified Iraqis to the left and right. On my second night at our basecamp, I watched a sentry aim their rifle at a young Iraqi

---

[13] 1st Marine Division, "Points from Security and Stability Operations Conference" (unpublished conference proceedings, 2002).

girl who was probably no more than five years old. She ran away screaming and the soldier laughed. I met her father later that week and apologized to him. He returned a forced smile.

But just like the Marine units, the Army had its ad hoc cultural experts. I spent time with a captain who took me out on patrols to meet with locals outside of our gates. He had a rural upbringing and saw little difference between his family and the people he met with in the farmlands around ar-Ramadi. The Iraqis loved him and were sorry to see him leave. Frankly, as an infantry officer with no cultural training, he was probably a better foreign area officer than I could have been at that point in my career. He had the right mindset. He was patient, humble, respectful, and genuine. Being good to people came naturally to him. That behavior paid dividends.

Unfortunately, this captain was one of a few exceptions in what we perceived to be an overly aggressive unit. We were quite happy to see them leave so we could stabilize al-Anbar. Most of us on the division staff believed that we would have a relatively easy time reducing tensions in al-Anbar Province. All we had to do was to treat people with respect and be Marines.

That optimism lasted about a month. It turned out that was our honeymoon period with the Iraqis. They wanted to see what we were about. Once they figured us out, the violence escalated. Smiling and waving at people did not stop them from shooting at us. In fact, some of them would smile and wave back while they pressed "send" on their cellphones to detonate roadside bombs. Cultural courtesy alone did not translate into mission success.

## Learning and Adapting in ar-Ramadi
Along with a few other members on the staff, I set about trying to understand the tribal networks in al-Anbar to engage with leaders and ease tensions. We received large folders of intelligence information implicating some tribal elders in violence. We received almost no information on the relative in-

fluence of various groups or leaders. We had to learn the hard way that most tribal leaders just played tribal ombudsman between 2003 and 2005. A few directly supported or led insurgent groups.[14]

Intelligence soldiers from the departing Army division did tell us that there was one "bad" tribe in al-Anbar. They recommended that we focus on dismantling it and arresting its leaders. I started to meet with the tribe's leaders. During the next few months, I gradually gained an appreciation for the complexity of tribal influence, tribal relations, and the realities of Iraqi life.

Half armed with what I thought was newfound wisdom on Iraqi culture, I brokered a peace deal between what I assumed were two warring tribes. I had been told the so-called bad tribe had a long-standing feud with a so-called good tribe. If we could put that feud to rest, we might win both tribes over to our side. In fact, I was being deftly manipulated by men who recognized my lack of knowledge and experience. I unwittingly set up a sham truce based on faulty information.

We did have some real success, though most of it based on work done by infantry Marines and soldiers in rural areas far from the spotlight. My daily engagements led me to discover that the provincial chief of police was working for the insurgents. Our follow-on evidence gathering and his arrest were only possible because I spent hours drinking tea with tribal leaders. It is quite likely they were manipulating me to take out a rival, but the end result benefited us both.

During seven months in ar-Ramadi, I gradually learned what I should have learned in my foreign area officer training and education. I came to understand the real-world complexi-

---

[14] CWO 4 Timothy S. McWilliams and LtCol Kurtis P. Wheeler, eds., *The Anbar Awakening*, vol. 1, *American Perspectives: U.S. Marines and Counterinsurgency in Iraq, 2004–2009* (Quantico, VA: Marine Corps University Press, 2009).

ty of individual identity, of group dynamics, and of the power of culture to influence human behavior.

In January 2004, when I looked at the province map and saw blocks of color representing tribal boundaries, I took them at face value. I dutifully memorized tribal names and hierarchies from our intelligence files. By August 2004, I knew all of that information was not only partly or mostly wrong, it was also dangerously and misleadingly precise.

No tribe in al-Anbar controlled street boundaries in a city. Hierarchies were a fantasy. Titular tribal leaders were often figureheads. Every Iraqi valued their tribal identity differently. Some had fierce tribal loyalty. Others could not have cared less. Money had as much or more influence on loyalties than names. Iraqis switched allegiances to survive and to feed their families. Nothing was what it seemed, and we were lucky to see a sliver of reality on a given day.

Marines with little or no cultural training had an even harder time trying to understand what was happening in al-Anbar Province. Their frustrations mounted as their Iraqi security forces melted away under fire, as they were deceived, and as they watched their friends die for a people who clearly did not want them there. Empathy for Iraqis was hard to come by on Marine forward operating bases in mid-2004.

Late one night during my 2004 tour, I wrote "Marines Are from Mars, Iraqis Are from Venus" as a simple guide to help Marines put themselves into the Iraqis' shoes.[15] I was surprised when it became a widely used training tool for predeployment courses and training exercises. This represented another indicator of the depths of our collective ignorance even a full year into our war in Iraq. I returned home later that year to a widespread awakening. Culture was a hot topic and money

---

[15] Ben Connable, "Marines Are from Mars, Iraqis Are from Venus," *Small Wars Journal*, 30 May 2004.

had started to flow into the gaps in training, education, and resources.[16]

## Building a Cultural Training Capability (on a Base of Sand)

Shortly after returning home, I was asked to represent the military at a two-day academic workshop in Rhode Island.[17] I spent both days serving as a punching bag for social scientists who appeared to have little respect or regard for the military. This experience also showed me the limits of my own knowledge. I did not understand the academic process of critique. Further, I was unaware of the terrible history between the military and the social science community in the United States. As a result, I could not communicate the military's need to use culture to "reduce the necessity for the use of violence."

After two days of shellacking, I was in utter disbelief that social scientists would not jump at the chance to improve military cultural understanding. I was frustrated that the people best positioned to help us were so reluctant to do so. The most useful line I could come up with to convince them was, "If you don't help us, we're going to do it anyway." I was implying that we would inevitably screw it up without their help. That plea actually worked with a few but further alienated others. I left the workshop—the first of many such workshops and conferences—with a few new colleagues and a desire to improve my ability to understand the academic perspective.

In late 2004, Arthur Speyer and I started the Cultur-

---

[16] *Defense Language Transformation Roadmap* (Washington, DC: Department of Defense, 2005).

[17] "Prepared for Peace?: The Use and Abuse of 'Culture' in Military Simulations, Training, and Education" (workshop, cohosted by the Pell Center for International Relations and Public Policy at Salve Regina University and the Watson Center for International Studies and Public Affairs at Brown University, Newport, RI, 6–7 December 2004).

al Awareness Working Group.[18] Speyer was the head of the culture program at the Marine Corps Intelligence Activity (MCIA), a job I would colead with him from 2006 to 2007. We reached out to everyone interested in cultural training, education, and knowledge. Within months, we had representatives from across the military and academia sharing ideas and enthusiasm for improving military cultural practices.

Other experts started programs and groups in parallel. Together, we seized on the short-lived interest in culture to build something that might endure. Most of us were realists. While we got some institutional support to build cultural capabilities, culture was still a hard sell for many servicemembers. We knew interest would inevitably wane. Kerry Fosher best articulated our realism-anchored enthusiasm with the mantra, "We have to hide bits and pieces of cultural capability that will survive the inevitable loss of interest."

## Lessons for a Culture Instructor

In early 2005, I volunteered to support the brand-new Marine Corps Center for Advanced Operational Culture Learning (CAOCL, a particularly horrible acronym in a military culture known for bad acronyms). This second job, and yet another part-time instructor position, took me around the country to train military personnel on their way to Iraq. This task allowed me to speak with and listen to about 3,000 people from every Service.

My experience as a cultural knowledge instructor sobered me to the challenge of preparing large numbers of people with wildly diverse experiences for complex operations. At best, I had a few hours with each group before they shipped off to

---

[18] See endnotes in Vanessa M. Gezari, *The Tender Soldier: A True Story of War and Sacrifice* (New York: Simon and Schuster, 2013); and LtCol William D. Wunderle, USA, *Through the Lens of Cultural Awareness: A Primer for US Armed Forces Deploying to Arab and Middle Eastern Countries* (Fort Leavenworth, KS: Combat Studies Institute Press, 2006).

Iraq. Often, I represented their only predeployment cultural training.

Some recipients were plainly hostile. In one case, I briefed a deploying Marine division staff. Three general officers, including a legendary retired Medal of Honor recipient, stared daggers at me from the front row as I suggested ways to understand Iraqi culture. During their deployment to Iraq, the commanding general of that unit would tell one of my CAOCL colleagues, "We don't need any of this culture shit, this is a gunfight."[19]

Based on these experiences, I learned to include phrases like "understanding culture in order to win" and "cultural terrain" into my lesson plans. The terrain analogy drove my academic colleagues up the wall because it was oversimplified and scientifically inaccurate. But it was a good compromise term for skeptical servicemembers and it helped bridge the gap between academia and the military.

I rejected the idea of teaching the basics of Islam or business meeting dos and don'ts. Regurgitating the liturgy of generic cultural information was often counterproductive: few absorbed it, fewer understood it in context, and most took those classes as an opportunity to catch up on sleep.

Instead, I focused my instruction on building empathy for Iraqis. I tried to help students understand what it was like to be an Iraqi in postwar Iraq. Classes focused on the realities of life under a brutal dictator, followed by a life in a country riven by war and chaos. Imparting an understanding of human suffering and its influence on behavior was more important than

---

[19] This statement came to me secondhand by a colleague in 2005. I have no reason to doubt its accuracy, particularly given my experiences with this officer and his staff in 2005 and early 2006. In addition to running what amounted to a counterguerrilla operation—hunting and killing rather than focusing on winning over the population—this officer's deputy gave me a direct order to omit all analysis from my analytic reports to avoid raising bureaucratic questions about U.S. operations.

passing along the gross overgeneralizations and racist tropes so many servicemembers had been exposed to by reading Raphael Patai's *The Arab Mind*.[20] Along with my colleagues, I spent quite a bit of time trying to undo the insidious effects of Patai's book, which had topped the military's recommended reading list for deploying servicemembers.

## Cultural Intelligence: Iraq, 2005–6

In December 2005, I redeployed to Iraq as the senior intelligence analyst in al-Anbar Province. My job was to collect all of the analyses for the province and write a daily narrative to help military and political leaders understand the course of events. For the first few months of my deployment, I worked for the general who did not need the "cultural shit" that a few of us thought might be important. Not only had the U.S. military failed to improve intelligence collection on cultural issues, it had moved in the opposite direction.

By early 2006, the Marine Corps intelligence organization had been completely subsumed by the high-value targeting kill chain. Intelligence systems collected information on the whereabouts of insurgents and fed that information to aerial or ground strike teams. I had several high-spirited arguments with some of my intelligence colleagues about the value of information. At the time, they held a steadfast belief that killing would solve our problems in Iraq. I told them that if we did not understand the social and political context of our actions that we might actually do more harm. Moreover, we might succeed by changing peoples' minds rather than killing them. This was, essentially, an argument more about the U.S. military mindset than about the nature of the information.

---

[20] Raphael Patai, *The Arab Mind* (New York: Scribner, 1973). Originally listed on the 2009 version of the Commandant's Professional Reading List, Gen James F. Amos added it to the 2011 list.

I lost that argument. In early 2006, the leaders of a nascent Iraqi movement to eject al-Qaeda from al-Anbar Province were assassinated when we did nothing to build on the movement or protect its leadership. In parallel, our own killing machine drove on. Insurgent violence in al-Anbar escalated despite our best efforts to kill our way to victory. As I departed al-Anbar in mid-2006, we effectively lost control of ar-Ramadi, the provincial capital. Colonel Peter Devlin, the head Marine intelligence officer, asked me to turn a slide deck on the situation in al-Anbar into a paper.

In the resulting 2006 *State of the Insurgency* paper, we argued that the situation had deteriorated to the point that there was no longer a military solution in al-Anbar.[21] We had lost the battle for influence over the population in large part because all of our resources were drawn away from analyzing the people of Iraq, ostensibly the focal point for any counterinsurgency operation. Iraqis had to reach a collective point of exhaustion with al-Qaeda before they would flip in 2007.

## The Human Terrain System and the Death Knell for Organized Culture Programs

Late in 2006, I started work at MCIA, leading the cultural intelligence program with Art Speyer. *Cultural intelligence* was an informal catchall for anything people related that did not fit into the normal intelligence kill chain.[22] As with cultural training and education, our small group of culture-focused intelligence experts faced an uphill battle. But we were making some incremental progress.

---

[21] I Marine Expeditionary Force, "State of the Insurgency in al Anbar" (declassified unpublished intelligence report, ar-Ramadi, Iraq, 2006). This document was cleared for open publication by the DOD's Office of Security Review on 16 December 2010.

[22] Ben Connable, *Military Intelligence Fusion for Complex Operations: A New Paradigm* (Santa Monica, CA: Rand, 2012).

Our biggest challenge came from competing proponents of cultural knowledge. Dr. Montgomery McFate was part of that early small circle of colleagues pushing for improvement. She was a heartfelt advocate for military cultural training and education. From 2003 to 2007, we spoke on panels together and built a friendship based on our common interests. But, during that same period, she helped take the military's cultural investments in a different direction.[23]

Instead of improving general cultural training and education across the military, the Human Terrain System (HTS) intended to provide tailored cultural expertise to military staffs to help them solve immediate tactical problems. HTS presented a tacit—and sometimes explicit—argument that across-the-board improvements were not possible. Instead of cultural training for soldiers and officers, HTS fielded teams of contractors with laptops to generate advice and reports for brigade staffs.

One of the sales pitch quotes for the program clearly conveyed the impossibility of general cultural training and education. An Army officer speaking about Iraqis said, "We don't ask them about their needs—paratroopers just don't think that way."[24] In other words, soldiers (and certainly aggressive airborne soldiers) did not have the mindset for nonviolent human interactions. The HTS approach was predicated on the belief that this kill-kill-kill mindset was immutable.

This was a powerful argument in the Pentagon. There appeared to be no clear or effective solutions to the problem of cultural training and education. Nobody could translate culture into the kinds of metrics that generally drove programming and investment in the Department of Defense. HTS offered up a concrete solution with a price tag and metrics.

---

[23] McFate contributed significantly to the publication of *Counterinsurgency*, Field Manual 3-24 (Washington DC: Department of the Army, 2006); and *Intelligence*, Joint Publication 2.0 (Washington, DC: Joint Chiefs of Staff, 2013).

[24] Connable, "All Our Eggs in a Broken Basket," 57–64.

It represented a neatly packaged answer to a nebulous and frustrating problem.

Art Speyer and I took up the fight against HTS from 2006 to 2008. We argued that low-level, across-the-board investments in cultural training, education, and intelligence would help to shift mindsets enough to alleviate the problems that had disrupted operations in Iraq and Afghanistan. Across-the-board investments would be sustainable over time, while HTS would be unsustainable. Investing in a contracted capability would leave us without a cultural capability as the wars in Iraq and Afghanistan wound down.

I pressed this argument home as the Marine Corps' representative to the Department of Defense board tasked with investing hundreds of millions of dollars to address the gap in cultural capability. At that time, Army general David H. Petraeus was a proponent of HTS. There were no powerful advocates for cultural training and education. Within a few years, the U.S. military would invest approximately $800 million in HTS.[25] This was effectively a zero-sum trade-off: that huge sum did not go toward sustainable cultural training, education, or intelligence.[26]

In 2009, I published a formal argument against HTS in *Military Review*.[27] This article served as a final quixotic broadside as I retired from the Marine Corps. "All Our Eggs in a Bro-

---

[25] Clifton Green, "Turnaround: The Untold Story of the Human Terrain System," *Joint Force Quarterly*, no. 78 (July 2015): 61–69; and Tom Vanden Brook, "$725m Program Army 'Killed' Found Alive, Growing," *Army Times*, 9 March 2016.

[26] Collectively, the Marine leaders pushing back against HTS managed to prevent it from becoming a Marine Corps program of record. In turn, this failure to win Marine Corps support undermined HTS program efforts to gain long-standing Joint force recognition and funding. Our successful fight contributed to the program's eventual demise.

[27] Connable, "All Our Eggs in a Broken Basket." See also Maj Ben Connable, "Human Terrain System Is Dead, Long Live . . . What?," *Military Review* (January–February 2018): 24–33.

ken Basket" earned me persona non grata status at the Army Training and Doctrine Command (the HTS parent organization) and not much else. At the same time, investment in cultural training and education stagnated. By the early 2010s, both HTS and the U.S. military's interest in culture were on their predicted glidepaths to nonexistence.

## Stashed Assets and Altered Mindsets

I drafted this chapter in 2020. Despite all of our setbacks, failures, and lessons unlearned, I think we can still claim success—we had low expectations. I do not think any of us expected to change the mindset of the entire U.S. military to allow for across-the-board agility in complex culture-centric operations. Instead, we held on to Kerry Fosher's directive to hide away some capabilities for the next Iraq or Afghanistan.

In 2020, HTS is no longer an active program. CAOCL has been canceled. The Army's formal culture programs are barely alive. Cultural intelligence never caught on. But on closer examination, culture is woven into lesson plans and educational curricula across the armed forces. Social scientists are positioned deep within the military bureaucracy, fighting a low-level insurgency to keep culture alive.

More substantial wins occurred in the minds of the servicemembers who have had years to sit back and think about their experiences in Iraq and Afghanistan. The kill-chain officers I argued with in al-Anbar Province in 2006 are now strong proponents of cultural training and education. They recognize that killing does not win irregular wars.

Culture is, quite loosely, human influence on human behavior. People set behavioral examples and also coerce and incentivize others to take on a certain mindset and to behave accordingly. Our struggle to improve cultural training has primarily been a struggle over the culture of the U.S. military. Our small group of advocates could never reshape the mindset of millions of people, but we could and did help many of them to interpret their experiences in war. We could and did help many of them prioritize culture in their efforts to understand war.[28]

I found the process of understanding and influencing mindsets to be far more important than memorizing cultural facts, avoiding cross-cultural friction, or even collecting the right data to inform military operations. If we can dispassionately empathize with the people we are working with—and those we are working against—the solutions to practical cultural challenges will be far less daunting. All of my experience and education led me to the conviction that military cultural training and education should primarily focus on developing understanding of the human condition not on the condition of any particular group of humans.

---

[28] For more on these concepts, see William Rosenau, *Acknowledging Limits: Police Advisors and Counterinsurgency in Afghanistan* (Quantico, VA: CNA and Marine Corps University Press, 2011); Norman Cigar and Stephanie E. Kramer, *Al-Qaida after Ten Years of War: A Global Perspective of Successes, Failures, and Prospects* (Quantico, VA: Minerva Initiative, Marine Corps University Press, 2011); Carroll Connelley and Paolo Tripodi, eds., *Aspects of Leadership: Ethics, Law, and Spirituality* (Quantico, VA: Lejeune Leadership Institute, Marine Corps University Press, 2012); Paula Holmes-Eber and Maj Marcus J. Mainz, *Case Studies in Operational Culture* (Quantico, VA: Marine Corps University Press, 2014); and Paolo G. Tripodi and Kelly Frushour, eds., *Marines at War: Stories from Afghanistan and Iraq* (Quantico, VA: Marine Corps University Press, 2016).

CHAPTER TWO

# On Becoming "Wise in the Ways of Others"
## Lessons Learned from Integrating Culture into Professional Military Education Curriculum

*by Lauren Mackenzie, PhD*

## Introduction

Although it is certainly not a new revelation, one of the driving forces behind my motivation to be part of the culture efforts in the Department of Defense (DOD) was to help military students—regardless of learning level—improve the quality of their intercultural interactions. As Mr. George Dallas has reiterated throughout his tenure as the Marine Corps culture center director (see chapter 9), culture training and education are always about *people*. At any given time, military personnel are preparing to work alongside, among, or against people who often look at and act in the world very differently than they do—and my teaching has been an attempt to offer tools for them to anticipate and manage those differences.

Admittedly, I have remained fairly far removed from the policy and strategy decisions made in the early years by Brian Selmeski, Kerry Fosher, Allison Abbe, and others, so I will not comment on that aspect of the DOD culture effort. The area I have been heavily involved with is instruction and have delivered culture-based presentations ranging from large lectures to small seminars every month annually since 2009. These classes have provided ample opportunity for me to make mistakes, attempt to learn from them, and update my working knowledge of best practices for integrating culture in the various learning levels across the professional military education (PME) spectrum.[1] Although PME is only one small slice of the larger culture efforts, it is the one in which I have had sustained involvement across branches of Service (e.g., Army, Air Force, and Marine Corps), and would like to expand on some of my experiences in the hopes that others might be able to use them should the DOD consider another turn toward culture.

The remainder of this chapter describes how I attempted to create a role for myself as a culture educator for PME students. I offer examples of lessons learned (and some favorite quotations that help capture these lessons) gleaned from experiences across different branches of the Services, learning levels, and teaching modalities. First, I emphasize the challenges I experienced with common premises associated with culture education; next, I offer a specific example of a tool I used to try to address this challenge; and finally, I conclude with some recommendations for those who find themselves charged with the task of integrating culture content into PME curriculum.

## "How You See the Problem May Be the Problem": 
*An Attempt to Offer a Different Lens*[2]
Efforts to support students as they enter different cultural contexts with the goal of interacting in culturally appropriate ways dates back to work by Edward T. Hall and associates

---

[1] This includes both enlisted and officer education ranging from the Sergeants School through the Marine Corps War College (colonel level).
[2] Lance M. Bacon, "Commandant Looks to 'Disruptive Thinkers' to Fix Corps' Problems," *Marine Corps Times*, 4 March 2016.

at the Foreign Service Institute in the 1940s.[3] The focus on micro-level cultural behaviors led to theories of intercultural communication that have since continued to evolve. There is no question that my background in the field of communication has strongly impacted the way I think about, talk about, and research culture, which, for better or worse, has differed from somewhat more traditional approaches gleaned from international relations or cross-cultural psychology.[4] What I mean by this is that every decision I have made in my culture classes is grounded in the question: How can this content be used to help students improve the quality of their communication? The assumption that *we do not interact with cultures, we interact with people* has framed my approach to course development and delivery.

With this assumption in mind as I began work for different DOD culture centers, I wanted to emphasize the kinds of communication practices that had the potential to transform students' perceptions surrounding the utility of culture education. I have attempted over the years to answer this question by building on what military students find problematic about current culture education. This process began by examining students'—and, for the last six years, *Marines'*—premises about culture and continued with my own realization of the

---

[3] Wendy Leeds-Hurwitz, "Notes in the History of Intercultural Communication: The Foreign Service Institute and the Mandate for Intercultural Training," *Quarterly Journal of Speech* 76, no. 3 (1990): 262–81, https://doi.org/10.1080/00335639009383919; Everett M. Rogers, "The Extensions of Men: The Correspondence of Marshall McLuhan and Edward T. Hall," *Mass Communication and Society* 3, no. 1 (2000): 117–35, https://doi.org/10.1207/S15327825MCS0301_06; and Everett M. Rogers, William B. Hart, and Yoshitaka Miike, "Edward T. Hall and the History of Intercultural Communication: The United States and Japan," *Keio Communication Review*, no. 24 (2002): 3–26.

[4] The main distinction between these fields of study lies in the primary level of analysis. Whereas international relations scholars emphasize the study of *institutions* and psychologists focus on the *individual*, a communication perspective takes *interaction* as its primary theoretical and practical concern.

ways certain aspects of the Corps' culture impact Marines' perceptions about culture education.

The past decade has taught me that learning about military culture from Marines (or soldiers, or airmen, etc.) themselves is necessary in order for them to have a meaningful educational experience about culture from me. Both formal and informal conversations with Marines, in the classroom and outside it, challenged my assumption that the value of culture education (which I took as a given) would be similarly apparent to my students. Before becoming a PME faculty member, I had taught at four different state universities in communication departments where my culture classes were either general education classes—and, as a result, often had long waitlists to get into—or upper-level electives that also made them desirable for students since they were one of the few small seminars students could enroll in as an upperclassman after years of the large lecture classes common for introductory courses at state schools. I had, therefore, never been in the position of having to "sell" my classes to students. As I began teaching at Air University at Maxwell Air Force Base in Montgomery, Alabama, in 2009 and Marine Corps University in 2015, I made the mistake of assuming time and again that PME students would be as drawn to my classes (which, looking back, had very little to do with me) and thus I did not spend enough time contemplating ways that I could make my culture classes relevant and useful for students who already had a full plate of courses to complete and schoolhouses who already had a full cadre of faculty to teach them.[5]

Thus, I moved away from focusing solely on the culture content and toward a focus on local perceptions about the value of culture education itself. To find the best ways to reframe

---

[5] In my current position as an academic chair at Marine Corps University (MCU), my work entails integrating culture content across the MCU schoolhouses, which includes the enlisted college and the four colleges devoted to officer education.

and communicate course content in a way that was useful for students, I adapted my discourse regarding its utility. In doing so, my hope was that the students would begin to recognize the value of culture education as providing communication strategies for managing the kind of challenges inherent to interaction with diverse groups. This communication lens was problematic in some ways initially, though, because I did not have the requisite language to connect to students' military experiences. Culture, like communication, is so broad that it can simultaneously mean everything and nothing (not to mention being considered soft and squishy) for students focused on a narrow definition of warfighting. This led to another challenge I had to contend with: the common presumption that since culture and communication are things we are immersed in all day every day, *anyone* can teach it.

## Just Because You Have Teeth Does Not Mean You Can Be a Dentist:
*Understanding Cultural Premises*

Working with a wide range of military students has taught me that multiple deployments and/or overseas assignments were often more the norm than the exception for most and this provided many students with exposure to various levels of culture training and education throughout their careers. Over time, I have noticed several premises that subsume how culture is described by many military students.[6] I developed an appreciation for the values underlying these premises through interviews, classroom discussions with students, and informal conversations with active duty and retired servicemembers at

---

[6] Kerry Fosher, "Cautionary Tales from the US Department of Defense's Pursuit of Cultural Expertise," in *Cultural Awareness in the Military: Developments and Implications for Future Humanitarian Cooperation*, ed. Robert Albro and Bill Ivey (London: Palgrave Macmillan, 2014), 15–29, https://doi.org/10.1057/9781137409423.0005; and Paula Holmes-Eber, *Culture in Conflict: Irregular Warfare, Culture Policy, and the Marine Corps* (Stanford, CA: Stanford University Press, 2014).

both the Air Force and Marine Corps culture centers. These conversations were supplemented by a thorough review of the literature devoted to military students and culture education. The result of such efforts revealed several cultural premises that presented barriers to me in the early years of my teaching:

- Culturally educated individuals are considered outsiders (e.g., role players, interpreters, embassy personnel, etc.) and therefore may not be familiar with how to make culture applicable to and appropriate for military operations.
- Conversely, anyone can do it. If you have deployed or lived overseas, you are a qualified culture educator. This presents challenges in that such individuals often lack the depth of knowledge necessary for effective postsecondary education.
- Culture as a "thing" that is not useful. It is often described as a "framework" or a thing that can be "sprinkled" on servicemembers before they deploy in an effort to provide "just-in-time" information.
- Culture as training and not education. It is referred to as a list of "dos and don'ts" or synonymous with "etiquette training" that is not in line with PME.
- Culture as burdensome. It is viewed as "death by PowerPoint," just another "check the box" requirement and, very commonly, another "rock in the pack."

While I wanted to use an understanding of these premises to inform my teaching, I also knew I needed to review some of the research that moved from my anecdotal observed patterns and toward a more systematic investigation of Marines' perceptions of culture instruction. For example, Paula Holmes-Eber, Erika Tarzi, and Basema Maki analyzed a survey completed by more than 2,000 Marines to identify the factors that influence their perception of the value of culture on military operations.[7] Their findings reflect that Marines' level of education, combined with a multicultural background, travel experience, and frequency of interaction with the local population during a previous deployment predicted Marines' attitudes about the importance of culture. They found that deployment alone was not a predictor but that satisfaction with the quality of culture training/education was. This led me to give more thought to how Marines' own premises about the value of culture education may enable an adaptation of both its content and perception. The literature pertaining to best practices for teaching military students suggests that they, like most adult learners, have a desire for immediate applicability and learn best when confronted with a particular problem that needs solving.[8] I have thus adapted many of my instructional processes to promote a discourse about a transferable mindset and skill set (i.e., skills that can be useful regardless of cultural context) rather than a memorized, region-specific set of dos and don'ts.

Another challenge with teaching culture in the DOD can be characterized as a tension between generalizable and specific knowledge. That is, to understand how military students perceive the value of culture education, it is important to rec-

---

[7] Paula Holmes-Eber, Erika Tarzi, and Basema Maki, "U.S. Marines' Attitudes Regarding Cross-Cultural Capabilities in Military Operations," *Armed Forces and Society* 42, no. 4 (2016): 741–51, https://doi.org/10.1177/0095327X15618654.

[8] Bradley Carter, "No 'Holidays from History': Adult Learning, Professional Military Education, and Teaching History," in *Military Culture and Education*, ed. Douglas Higbee (London: Routledge, 2010), 167–82, https://doi.org/10.4324/9781315595405; and Susan Steen, Lauren Mackenzie, and Barton Buechner, "Incorporating Cosmopolitan Communication into Diverse Teaching and Training Contexts: Considerations from Our Work with Military Students and Veterans," in *Handbook of Communication Training: A Best Practices Framework for Assessing and Developing Competence*, ed. J. D. Wallace and Dennis Becker (New York: Routledge, 2018), 401–13.

ognize their need for knowledge that is directly and immediately useful to where they think they will deploy next. Similar to what Edward Hall and his associates found in the 1940s, culture-specific training is the appropriate means to that end.[9] However, the unpredictable nature of military operations calls for a generalizable set of skills for thinking and interacting that can be used flexibly in an attempt to prepare students for the unknown.[10] Such skills include the ability to ask better questions, use multiple lenses to examine a problem, and check perceptions—all of which can be thought of as universally applicable and useful beyond an individual mission or deployment. Work toward developing a growth mindset and broadening military students' perceptions about culture education as a skill set that is applicable across contexts is where my attempts at transformation began to take shape.[11]

## Just Because Something Is Common Sense Does Not Mean It Is Common Practice:
*An Attempt at Reframing through Broader Applicability*

Transformation has been described as changing one's mental picture of a problem and illuminating it in a different way.[12] For educators, such transformation often involves an adaptation of the ways in which students, instructors, and content interact. Conversations with active duty and retired military personnel, instructors and students, contractors, and military and academic curriculum writers have added to my "living" repository of the ways culture is described. Such conversations

---

[9] Leeds-Hurwitz, "Notes in the History of Intercultural Communication," 262–81.
[10] Brian R. Selmeski, "Military Cross-Cultural Competence: Core Concepts and Individual Development," in *Armed Forces* (Kingston, ON: Centre for Security, Armed Forces and Society, Royal Military College of Canada, 2007).
[11] Carol S. Dweck, *Mindset: The New Psychology of Success* (New York: Random House, 2006).
[12] Linda L. Putnam and Samantha Rae Powers, "Developing Negotiation Competencies," in *Communication Competence*, ed. Annegret F. Hannawa and Brian H. Spitzberg (Boston, MA: DeGruyter Mouton, 2015), 367–95.

have also highlighted, however, that there can be a glaring disconnect between knowing something and deliberately putting such knowledge into practice. I have been continually reminded over the years that I am not just advocating for the value of intercultural communication content in the curriculum, but I also need to embody such concepts and skills in these conversations. This realization played a role in the two-way process of cultural adaptation on my part, where I treated education as interactions with students that jointly produce knowledge, as opposed to a one-way delivery of content.[13] So what does this look and sound like?

One instance that reflects how I attempted to transform a perspective of culture education from problematic to useful was through the cocreation of the *Why Culture?* video series.[14] The goal of the videos was to try to alter the way culture content is framed and valued by featuring both an academic overview of certain culture concepts as well as Marines themselves talking about the value of such understanding in military operations. As is the case with most students, the "who," the "how," and the "why" matter to Marines. The three videos in the series (each approximately five to seven minutes long): *Why Culture?*, *The Value of Culture General*, and *The Foundational Cross-Cultural Competence Skill of Perspective-Taking* follow a similar logic.[15] The thinking that informed the creation of the culture video series was that a credentialed academic (who)

---

[13] Jo Sprague, "Why Teaching Works: The Transformative Power of Pedagogical Communication," *Communication Education* 42, no. 4 (1993): 349–66, https://doi.org/10.1080/03634529309378951.

[14] *Why Culture?* (Quantico, VA: Center for Advanced Operational Culture Learning, Marine Corps University, 2017), 5:20 min.

[15] *Culture-general* is the approach taken by the Marine Corps to advance generalizable culture concepts and skills that are transferable across contexts. Culture-general content (i.e., holism and identity or suspending judgment and perspective-taking) is offered to Marines to make culture-specific learning more efficient. Kerry Fosher et al., *Culture General Guidebook for Military Professionals* (Quantico, VA: Center for Advanced Operational Culture Learning, Marine Corps University, 2017).

would provide some culture content, an influential Marine would explain its usefulness (why), and that it would be offered in a short video (how) that is accessible to Marines in a classroom or on their smartphones via YouTube while waiting for a flight, thus avoiding to some extent another "rock in the pack" of more online training requirements. To create this series, several formal and informal conversations were undertaken with Marines (retired and active duty) to determine the most appropriate content to deliver and the most accessible way to deliver it.

To illustrate how Marine perspectives and voices are incorporated into the series, one video begins with the voice of General James M. Mattis, former secretary of defense, providing a rationale for *Why Culture?*:

> If Marines want to be prepared for the next fight, they are going to have to prepare themselves to adjust rapidly to whatever culture they're going into. We are always operating in other people's cultures. We've got to be able to adapt to those cultures and make certain that we do not create problems.[16]

The intent was to begin the video with General Mattis in order to emphasize the importance of culture to mission effectiveness through the words of an influential Marine articulating what should be a shared value. There are two key points about his statement that relate to cultural premises. First, Mattis's specific statement about being sure "we do not create problems" is well understood by Marines who have watched themselves become the focus of national news that publicized instances where military personnel made costly mistakes due to varying levels of cultural misunderstanding or what some cultural groups felt were actions that did not respect other

---

[16] *Why Culture?*.

cultural values.¹⁷ The point here is that intercultural interactions Marines have while overseas tend to be both high stakes and highly visible. Second, Mattis used Marines' own growing awareness of mistakes to make a strong point about the importance of Marines not only being able to accomplish their mission but also about understanding the culture of the people where and with whom they will be working. In this way, he articulated culture as place-bound, and context-specific. It was my hope that Mattis's articulation of this point would resonate with Marines, many of whom are accustomed to carrying their own culture while often needing to temporarily adjust to others.

Although General Mattis offers a powerful voice that has the potential to lend credibility to the "why" of culture, I also learned that Marines want to hear the "how" from their peers as well. Several minutes into the *Why Culture?* video, a junior Marine tells the story of how an understanding of a cultural concept devoted to kinship worked for them while they were deployed:

> I've dealt with more local nationals than anyone else during my deployment because I've had to deal with them as convoy escorts, and for operating equipment, and other things. In Jordan, we were working on contracts. And I learned that military rank was less important than someone's last name, the family they came from. In culture general, you can talk about informal and formal power structures. How I define that—the formal structures are "who" as tasking authority. But the informal is who has influence on the people around you.

---

¹⁷ A well-publicized instance of this was the 2012 burning of Korans by U.S. military personnel, summarized in Craig Whitlock, "U.S. Troops Tried to Burn 500 Korans in Blunder, Investigative Report Says," *Washington Post*, 27 August 2012.

> Over there, they called that wasta. So it's important to know about that, because in my job, I need to know who can talk to whom. I got a clearer site picture about all that when I was over there, but I'd still say it was ambiguous sometimes.[18]

This more specific example is an important step toward portraying culture not as a "thing" or list of dos and don'ts, but as a way of thinking more holistically about relationships. Marines, like most students, relate well not only to narrative but to specifics about what they need to be aware of to do their jobs better. A key goal with the creation of the video series was to emphasize the point that the "general and specific," "training and education," or "online and in-person" dialectics do not need to be thought of as in opposition if they are framed instead as enablers providing students the tools needed to be better Marines.

I have continued to try to involve Marines in the ongoing creation of their own culture education so as to render the content relevant across contexts and interpersonal interactions. This work has involved designing instructional interventions using the "who" (credentialed academic and Marine voices), "what" (culture-general concepts/skills accompanied by military-relevant examples), "why" (provided by an influential Marine), and "how" (in the form of accessible anytime, anyplace media) logic that articulate Marines' evolving perspectives to eventually adapt the way culture education is practiced over time. Engaging students in conversations where they reflect on the culture courses designed to help them become more effective in their jobs is not simply an example of

---

[18] *Anonymized Interview from Longitudinal Assessment Project*, Translational Research Group Report, IRB Protocol #USMC.2016.0005 (Quantico, VA: Center for Advanced Operational Culture Learning, Marine Corps University, 2016).

communication best practices in distance education, but it is also an example of putting intercultural communication theory into practice in PME.[19]

## Conclusions:
### Research as Me-search

Along with the broader considerations mentioned above for making culture relevant and accessible for military students, I will conclude with two specific recommendations for those who may find themselves tasked with integrating new or different content into existing PME curriculum. First, try whenever possible to position yourself as support. The schoolhouses often have a full roster of core faculty who feel they could teach culture classes (delivering a large lecture or facilitating a seminar) just as well as anyone else. Make the time to develop a basic understanding of: (1) the existing curriculum, (2) what each faculty member's academic background is, and then (3) formulate an elevator pitch explanation of how you add value. Although it took me years at Marine Corps University, I stood back and waited to find a gap in the curriculum that I felt an intercultural communication approach could fill. I advocated for the value of piloting courses devoted to "Metacognition" and "The Impact of Culture on Critical and Creative Thinking," for example, once I read through various publications put forth by senior leaders that called for such competencies among military students. I have made it a point over the years to articulate to faculty and leadership that I am not attempting to add another rock to the pack, but rather I am offering content that could enhance the quality of the courses that already exist. A common metaphor that captures this point is comparing the curriculum to a jar full of marbles. Although

---

[19] Lauren Mackenzie and Megan Wallace, "Intentional Design: Using Iterative Modification to Enhance Online Learning for Professional Cohorts," in *Communicating User Experience: Applying Local Strategies Research to Digital Media Design*, ed. Trudy Milburn (Lanham: Rowman & Littlefield, 2015), 155–82.

there is no space for another marble, there is often empty space between the marbles that could be better used.

The next recommendation (and I recognize this so-called "widening the aperture" phrase is fairly common) is to broaden the applicability of culture as much as possible. I have found it useful to frame culture concepts/skills as overlapping with areas military students might be more familiar with, such as critical thinking, leadership, and red teaming. I consistently make the point that no single academic discipline has the monopoly on dealing with difference or interoperability, but rather there are a constellation of overlapping competencies that, taken together, can offer important insights for students interested in managing the challenges associated with cultural complexity.

Although the examples in this chapter offer lessons learned at an individual level, there are certainly implications for broader consideration. One key challenge has been to provide metacultural education for military students that works (in potentially high visibility positions) all over the world. Rather than teaching the specifics about a particular culture group, my instruction has aimed to convey to students the value of generalizable and transferable interaction skills that will help them become more culturally appropriate communicators. Reflecting on my time working for and alongside the military has reminded me how much commonality there is between my personal and professional interests. I am no exception to the characterization of research as "me-search" and am consistently looking for ways to improve my own communication practices (especially during difficult conversations), while I work to help students to improve their own. One of my earliest mistakes was not pushing back when war college directors would minimize the importance of culture and communication at the strategic level, reminding me that, as O5s (field grade officers or above), "they got it." Instead of accepting that as true, I wish I had come to such meetings and introductions with a sharper response that put into practice some of the recommendations I have summarized in this chapter. There was so much of my own communication course content I could have applied to make a stronger case for the value of culture education at any learning level. I had a laundry list of responses 10 minutes after these conversations, when it was often too late. Luckily, I got several more chances to formulate a more effective reply given that active duty school directors typically rotate out every two years.

I will be the first to admit that I still have plenty to learn from studying the cultural practices of PME students, and I have learned a great deal from examining my own assumptions and practices as a culture educator. It has been said that cultural researchers seek nothing more and nothing less than to "become wise in the ways of others."[20] Although I make no claims to have fully achieved such wisdom, years of being in the classroom with students who have challenged me to rethink my assumptions and reconsider my frames of reference about culture have given me an enduring appreciation for military learning. In education, as in life, our mistakes are often our best teachers.

---

[20] John J. Pauly, "A Beginner's Guide to Doing Qualitative Research in Mass Communication," *Journalism Monographs*, no. 125 (February 1991): 1–29.

# CHAPTER THREE

# From Aha Moments to Emerging Stories of the Good Old Days
## Reflections from Many Years in a Fascinating Field

*by Susan Steen, PhD*

## Introduction

If I had to pinpoint how my interest in culture education and training began, I would say that it occurred not as a singular "aha!" moment but as the culmination of a process stemming from my first immersion experience overseas, studying abroad during my junior year at university and returning home to discover, with no small surprise, that I lived in a culture too. Through exposure to a different country and people, I became more curious about my own, seeing elements of my home culture that had never registered as "my culture" previously. By way of the experience abroad and the reentry process, I had learned about myself and my own culture just as I had about the people and the culture of my study-abroad destination country, and it had all been eye-opening and life changing. I wanted to help others achieve similar revelations.

Consequently, and subsequently, I volunteered to help with my university's study-abroad and international student orientations. Following my undergraduate graduation, a master's degree, and another stint abroad as a Rotary International Ambassadorial Scholar, I took my first job as a professional international educator at the very university that sent me abroad in the first place. A few positions, a couple of universities, and more than 20 years (of teaching, training, and developing international and intercultural programming for students and faculty) later, I joined the Air Force Culture and Language Center (AFCLC) as the assistant professor of cross-cultural communication. I am in my fifth year with the AFCLC, a center embedded in the Air War College at Air University, at Maxwell Air Force Base in Montgomery, Alabama.

Although my "audience" has changed throughout the years—I am now teaching military professionals ranging from enlisted forces to general officers, whereas before I was largely working with civilian undergraduate students—my goals have not. Ultimately, I want to help our everyday citizens, our troops, and our leaders develop the mindsets and skill sets essential to effective communication in twenty-first century global contexts—a world of highly connected people and increasingly porous borders in which cybercrime, global disease, natural disasters, conflict, and humanitarian crises do not adhere to boundary lines drawn on a map, and successful multinational partnerships and coalitions are essential to confronting them. Additionally, I hope that my students learn to value and appreciate the similarities and differences among people, within and across different cultural groups. Within PME, my focus is on helping students cultivate skills to build effective relationships with our international partners and to develop strategic empathy so that they can better understand our allies and better analyze our adversaries.[1] Understanding different worldviews, learning to see things from others' perspectives, and discerning the "why" and not just the "what" of human behavior are critical capacities for our military leaders.

---

[1] For discussion and an extended example of strategic empathy employing a historical perspective, see H. R. McMaster, "How China Sees the World," *Atlantic*, May 2020.

Throughout my culture education career, I have used a communication perspective to frame my scholarship and instruction. A communication perspective examines how messages influence the ways people make meaning and act, taking into account both the *process* and the *effects* of messages. It considers communication—the means through which human interaction is conducted and relationships are forged, maintained, and transformed—a vital force in shaping social reality. It suggests that, more than just a vehicle for transmitting information, communication affects how we construct meaning, create shared practices, negotiate social reality, lead and follow, and develop our understanding of ourselves and our place in the world.

The communication perspective offers frameworks and approaches to the teaching of culture that are different from the (more common within the DOD) social psychology or anthropology perspectives, and it contributes richly to my work in the Air Force Culture and Language Center, where our faculty team of anthropologists, regional scholars, social historians, and me (the sole communication scholar), applies a robust interdisciplinary lens to our investigation of culture. My discipline focuses on the central role of communication in shaping our social worlds, and I examine culture from this standpoint because our communicative practices and patterns bear distinct vestiges of our cultures, even as our communication shapes and transforms our cultures. For example, I am currently teaching a course on creating cultures of resilience, drawing on communication theories that emphasize the socially constructed nature of sense-making and that underscore the role of agency in cultivating healthy communication practices and managing complex and often unpredictable interaction, especially following disruption or disaster. My military students have found these communication frameworks useful in a range of contexts: from disciplinary events, to cross-cultural negotiation, to engagement of moral injury and post-traumatic stress disorder (PTSD), to strengthening resilience in and across military units.

As I consider the impetus for this important and well-timed volume and reflect on my own work and those that will follow in PME culture education, three lessons present themselves. These span my professional history in both civilian and military contexts. The first is that at the grounded level of the classroom, students are (like I was) almost always surprised to discover that culture is not just something "others" have but that "we" do too. The second is that the necessity for culture education does not simply disappear even when DOD interest waxes and wanes with leadership changes or with budget constraints, so it behooves our PME culture experts to continually make evident and communicate the relevance of our work. The final lesson involves the importance of establishing and maintaining connections and relationships across a variety of domains.

## The First Imperative

A consistent and recurring pattern in my teaching and training over the years is the emerging self-awareness that occurs through culture exposure, education, and training. In this sense, I am speaking not only of awareness of the individual self, but also the increased understanding of oneself within one's historical, collective, sociocultural contexts—something Louise J. Rasmussen and Winston R. Sieck have described as "self in cultural context" in their multilayered model of military cross-cultural competence, and that is likewise featured in *DODI 5160.70 Competency-Based Framework*.[2] People belong to a variety of cultural communities, but they do not necessarily

---

[2] Louise J. Rasmussen and Winston R. Sieck, "Culture-General Competence: Evidence from a Cognitive Field Study of Professionals Who Work in Many Cultures," *International Journal of Intercultural Relations* 48 (September 2015): 87, https://doi.org/10.1016/j.ijintrel.2015.03.014.

awaken to or recognize their own until they are exposed to a different culture through travel, deployment, encounters with people from different regions or countries, and the like. As Marshall McLuhan and Quentin Fiore point out, fish know very little about water since they live in it; they have no other environment with which to compare it.[3] This is an apt metaphor for the notion that until someone is exposed to a different world than the one they inhabit, understanding their own world may be facile at best.

When referencing culture education, people tend to automatically assume this entails teaching and learning about some amorphous group ("them") in some far-flung region of the world. And, of course, it does. But as students learn about other cultural groups' worldviews and their ways of being and doing, they are often learning about their own, as well, perhaps for the first time. American students need to consider how their own histories, identities, worldviews, and values have shaped their assumptions, practices, and systems of governance, just as they learn about the ways other groups' institutions, beliefs, and practices reflect their distinctive worldviews and histories. In my civilian higher education work as well as my military teaching, these discoveries have occurred through exposure to other cultures and peoples (e.g., via a deployment or study abroad experience or friendship with an international student) as well as in the classroom, which provides academic frameworks and vocabulary to help people make sense of their intercultural encounters and experiences.

In DOD classrooms, these "lightbulb" moments often occur when discussing assumptions and values, nonverbal and verbal communication patterns, and conflict response styles of other cultural and regional groups, in comparison with our own. For example, my students invariably enjoy drawing on Geert H. Hofstede's culture dimensions to understand the

---

[3] Marshall McLuhan and Quentin Fiore, *War and Peace in the Global Village* (New York: Bantam, 1968).

ways that different groups view hierarchy and power or group obligations versus individual needs, and to consider how these assumptions are reflected in different communication practices.[4] It helps that students are assigned specific readings on American history, society, and culture, as well as works designed to help them understand other cultures and regions of the world (the Middle East, parts of Asia, and regions of Africa, etc.). It helps likewise that these military students typically have a rich variety of lived experiences in different countries and regions that they draw from in making connections to classroom constructs. Regardless of how it occurs, such "self-in-cultural-context" awareness is critical to the cultivation of cultural competence, and it typically features as an essential element in models of the same. This lesson should not be overlooked or neglected by PME culture educators, and we must continually find ways to underscore and highlight the importance of self-understanding in our teaching and training.

## The Second Imperative

The next lesson learned involves establishing, maintaining, and messaging the relevance of culture education and training across PME and the DOD enterprise. It is important to foster such understanding and buy-in, especially from those who must make the tough budget decisions that can result in the continuation or the cut of such programs. Demonstrating the importance of our work is easier said than done, and the notion may seem distasteful to some educators and academics. But in an era of competing priorities and an ever-changing interconnected world where new challenges are constantly emerging to vie for space and attention, I view this work as essential. The (virtual) Language, Regional Education, and Culture Sympo-

---

[4] Geert H. Hofstede, *Culture and Organizations: Software of the Mind* (New York: McGraw Hill, 1991); and Geert H. Hofstede, Gert Jan Hofstede, and Michael Minkov, *Culture and Organizations: Software of the Mind*, 3d ed. (New York: McGraw Hill, 2010).

sium hosted by the AFCLC in October 2020 featured a variety of senior military and civilian leaders, including the former secretary of the Air Force, Barbara M. Barrett, who touted the importance of culture education and training and provided examples from their own experiences in key command positions overseas or in working closely with international partners and allies.[5] Testimonials such as these both bolster and signal the importance of our work, and we must continuously find ways to engage leadership on its value.

In my own professional history, it has not been hard to persuade the people directly involved, such as students or faculty in programs abroad or international students commencing their first year of a U.S. university degree program, of the relevance of culture education and training, especially when they are excited about the opportunities afforded by such experiences. The same has been true for my DOD students, from enlisted personnel enrolled in an online class sharing experiences of deployment that illustrate course concepts, to American and international lieutenant colonels in seminars exploring nonverbal codes and intercultural conflict, and finally to general officers assuming key command positions overseas. The need for such teaching and training may be less apparent, or perhaps less compelling, to university administrators making difficult financial decisions in the face of rising costs and declining funds, or to a Department of Defense whose level of interest in culture education has been inconsistent over the years and whose budgets are likewise unpredictable from year to year.

We in the teaching trenches know that the need for culture education and cultural capabilities among DOD personnel does not simply disappear along with waning priorities, vanishing budgets, and the programs and personnel associated with them. Indeed, it is as important today as it has always

---

[5] See Jasmine Bourgeois, "AFCLC and Air University's First Virtual LREC Symposium Draws Thousands of People," Air University, 21 October 2020.

been. What has been missing, perhaps, is the ability to frame our work in ways that make its importance clear to leaders at different levels of the DOD. But we must perform the same cultural agility we attempt to engender in our students in conveying the relevance of culture education and training, regardless of what it is called—intercultural effectiveness, cross-cultural competence, cultural intelligence, or otherwise.[6] While the challenges and priorities of the day may change, the need for understanding culture does not: culture is our lens for examining human behavior, and understanding the human domain is a critical element of DOD readiness. Our work is every bit as important now to understanding organizational cultures, great power competition, regional tensions, and issues of diversity and inclusion as it was to teaching about cultural contexts related to counterinsurgency as the U.S. presence in Iraq and Afghanistan was increasing. If we can see this but our leaders cannot, we must find ways to continually and conspicuously connect the dots.

## The Third Imperative

The final message I would impart to the next generation of PME culture scholars is one that I believe is critical to our professional health and well-being, to the nature of our work, and to the culture education enterprise. This lesson is on the importance of creating connections and community within and across our culture centers, institutions, PME, the DOD, and our academic fields. Indeed, collaboration with civilian and military faculty within our own and across different scholarly disciplines is a core element of some PME faculty performance standards—certainly, it is in my own. Candidly, I am a better advocate for this than I am a practitioner. But the relationships

---

[6] For a summary of terms employed by various scholars and practitioners, see Brian R. Selmeski, *Military Cross-Cultural Competence: Core Concepts and Individual Development* (Kingston, ON: Centre for Security, Armed Forces, and Society, Royal Military College of Canada, 2007), 5.

and friendships that I have developed within my own center, and across Air University (through participation in cross-organizational activities such as the faculty senate, interdisciplinary involvement in a research task force, and presentations through forums sponsored by our Teaching and Learning Center) have been invaluable, insightful, and rewarding. The communities I have engaged through consistent participation in the Air Force Culture and Language Center's annual LREC symposium, and recently the emerging PME Faculty Consortium (November 2019 and October 2020) organized by Dr. Lauren Mackenzie of Marine Corps University, have helped me better understand the cultural context in which I work and teach and connected me with some founding leaders of the field. Maintaining membership in professional associations and attending or presenting at conferences in my academic field have enabled me to create connections that span military and civilian worlds and have led to enriching scholarly collaborations with people whose work has long inspired me. While this is a general lesson that could be usefully employed by anyone regardless of whether they are in the "culture education business" or in a different field, such involvement hearkens back to and reinforces, in the particular context in which I am writing, the second imperative: maintaining recency and relevance. The practice of deliberately cultivating such community and connections is something I recommend for future PME culture educators.

 I feel deeply fortunate for the myriad of professional opportunities and experiences I have had in culture education over the years, and I am especially grateful to be part of the PME community in this place and at this time. Such opportunities may be fleeting, and we cannot predict with any certainty when the bell may toll, as it were, for the nature and context of our own work. But while there is no guarantee that I will have the chance to continue with my work ad infinitum, I am grateful to have been able to contribute to developing a more culturally competent military and citizenry. I hope the next generation of culture scholars finds itself presented with similarly rewarding opportunities, and I dare to hope likewise that I might still be around to share stories of these golden "good old days" with them.

# CHAPTER FOUR

# Surfing the Sine Wave of Military Culture Education

*by Angelle Khachadoorian, PhD*

## Introduction

Imagine loving your academic discipline—anthropology—so much that you believe it can change the world. You dedicate years of your life to developing your knowledge and skills. You choose to teach this discipline—often dismissed as being an interesting but frivolous subject—because it is taken seriously by your students. They, like you, feel that anthropological knowledge can help to mitigate cross-group conflict and potentially save lives. Then imagine that your disciplinary peers have a fraught relationship with your employer, and you experience that tension in both subtly unexpected and explicitly acted ways, all while trying to do meaningful work. This is the reality of being an anthropologist teaching culture for a military culture center (MCC).

The relationship between the community of American anthropologists and the U.S. military is tense and tangled with ethical concerns. Debates about the ethics of anthropologists working for the military have raged for decades. The newer conflicts surfaced over concerns about an Army program, called the Human Terrain System (HTS), that sought to embed anthropologists into deployed Army units and reached a crescendo in the years after the invasion of Iraq, leading to significant disciplinary efforts to define ethical standards for this relationship.[1] I was not a part of the ethical debates, and my knowledge is secondhand. My brief reference here is to set the context in which I entered employment with the United States Air Force.

There were massive reverberations for all anthropologists who worked for the military, regardless of their actual role, including that of teaching faculty. Employment by the military seemed to label these anthropologists as morally suspect. I knew this background and these labels when I entered the Air Force Culture and Language Center (AFCLC) in 2011, but I made an informed choice based on three key decision points.

First, I had previously taught Air Force personnel—both enlisted members and United States Air Force Academy (USAFA) cadets—and I found them intellectually engaged and eager to explore cultural assumptions. Second, I knew that my role was going to be as a faculty member, and as such, I was not expected to undertake activities that were ethically dubious. In fact, I asked about this issue in-depth during my job interview. Last, I knew that anthropological knowledge, and an understanding of the fundamental characteristics of culture, could serve to prevent the kinds of misunderstandings that may well have fatal results for both American military personnel and for local people. There was, I felt, a level

---

[1] See Montgomery McFate and Janice H. Laurence, *Social Science Goes to War: The Human Terrain System in Iraq and Afghanistan* (London: Oxford University Press, 2015), https://doi.org/10.1093/acprof:oso/9780190216726.001.0001; James Peacock et al., *Final Report, November 4, 2007* (Arlington, VA: American Anthropological Association [AAA] Commission on the Engagement of Anthropology with the U.S. Security and Intelligence Communities, 2007); and Robert Albro et al., *Final Report on the Army's Human Terrain System Proof of Concept Program* (Arlington, VA: AAA Commission on the Engagement of Anthropology with the U.S. Security and Intelligence Communities, 2009).

of urgency to teaching this material. It was an application of anthropological thinking that was more practical than theoretical and contributing in this way has always been important to me.

The ethics debates have had a significant impact on my own thinking about anthropology, the military, and my personal and professional values. It is somewhat ironic that the HTS program—problematic in many ways—has had any impact on me as an anthropologist since, to me, the term *anthropologist* was rather freely applied to HTS. I have met several people who had worked for HTS, and only one was an actual anthropologist. This reminds me of an experience I had early in my teaching career for the Bureau of Indian Affairs. A student spoke about his tribe's recent experiences with an anthropologist attending tribal events then writing publicly about private matters, which surprised me since I could not imagine one of my colleagues being so inappropriate. I asked how members of the tribe knew this person was an anthropologist, and after some back and forth, we realized the person described had been a journalist, not an anthropologist. Unfortunately, the term anthropologist has become a sort of catch-all word for someone who invades community privacy by asking intrusive questions about culture.

I am an odd duck of an anthropologist, as teaching about culture for the military puts me in a small but impassioned flock.[2] The term *odd duck* reminds me of a saying that is often used to support assumptions: if it walks like a duck and quacks like a duck, it must be a duck. That is what being an anthropology professor for the Air Force sometimes feels like—there are many assumptions placed on me in terms of my motives, behavior, professional ethics, and goals.

Part of the difficulty of being at the middle of this conflicted "military + anthropology" equation is that assumptions

---

[2] Anna Simons, "On 'Military Anthropology'," *Parameters* 50, no. 3 (2020): 121–24.

can come from both sides of the relationship. If it feels like a portion of my anthropological peers assume my work supports covert activities (a term with specific meanings to the military but used more freely by the civilian world), it also can feel like military personnel think I view them as reactive, callous, or aggressive. It is a strange place to be between two culture groups that are suspicious of each other. Luckily, liminality is the normal state for an anthropologist.

## Professional Paths and Unbeaten Tracks

Prior to my employment by the AFCLC, I had taught Air Force personnel in other contexts: the Community College of the Air Force (CCAF) at Kirtland Air Force Base in Albuquerque and the United States Air Force Academy in Colorado Springs. The first college-level class that I ever taught was for enlisted Air Force personnel through CCAF a few years prior to 9/11. The enlisted students were active in class, committed to learning, and eager to apply theory to their lived experiences. It was validating as a beginning college instructor, passionate about the topic I taught, to have students who approached the class with a heightened level of commitment.

After the CCAF course, I taught different social sciences for almost a decade at the Southwestern Indian Polytechnic Institute (SIPI), a Bureau of Indian Affairs/Education community college in Albuquerque, New Mexico. I learned the norms of federal service while also teaching students who might not otherwise have been exposed to anthropological thinking and knowledge in a positive way. Historically, anthropologists had spent considerable time in Native American communities, but rarely had they taught the unique ways of asking, thinking, and analyzing that anthropology offers. My interest in anthropology grew out of my efforts at collecting life histories in my own Armenian community, and I can attest to the value of anthropology for maintaining marginalized histories and community memories that might otherwise be forgotten. I realized at SIPI that teaching the application of anthropological knowledge, skills, and tools could offer more to my students than just citing some interesting cross-cultural facts. Anthropology was anything but frivolous.

At the end of my time at SIPI, I spent a year as a distinguished visiting professor at the United States Air Force Academy, where I enjoyed the cadets' excitement for my classes. When I taught about global human rights dilemmas or explained topics likely to cause cross-cultural conflict, I saw anthropology as a potentially lifesaving skill set. The cadets did also; my classes were always full. The cadets I taught knew that soon after graduation, they were likely to be deployed to a country where they would encounter cultures, values, beliefs, and behaviors significantly different than their own. They wanted the tools that anthropology offered.

## Lessons Learned in Teaching at a Military Culture Center

I am deeply concerned at the demise of military culture centers. Shutting them down does not erase the fact that, regardless of the technology brought to bear, human beings are at the basis of all conflicts. There is important knowledge, practical experience, and organizational insight being lost as culture programs shut down, personnel are laid off, and records get archived. I am, however, optimistic as I visualize a future era when MCCs experience a resurgence and graduating anthropology PhDs give serious thought to teaching at one. Allow me to offer, in the spirit of balancing the perspectives of both the military and academia, my lessons learned: know your own values, know your student population, know your positioning, and know how to adapt.

## Know Yourself

Before you apply to be a professor at a military culture center, you need to know your own motivations, your professional standards, and your ethical boundaries. Understand why you are choosing this unique position because your role will somewhat resemble, but not totally mirror, that of a professor in academia. Know what tasks, topics, or activities you would unequivocally reject, what would keep you up at night, and especially know the ways that you want to do good work. I knew my personal and professional ethical boundaries prior to entering my position. I worked to see what was expected of me, and I looked for the situations where I could do the greatest good. I knew what I would—and will still—say no to. This is the best and most useful piece of advice that I would give to any young professional in any field and in any job. If you have a clear understanding of your own value system, you are both far less likely to violate your own norms and less likely to say yes to something that you will regret later.

My professional ethics mean that I approach all of my research with a concern and respect for the privacy, consent, and well-being of the people I work with. This is no different than if I were in a standard university setting rather than in professional military education (PME). I teach about culture in general terms, and I offer graduate-level seminars on social science topics, such as the intersection of culture and war. My focus is to provide senior military officers with tools and perspectives for understanding cultural communities that are very different than their own as well as offer a means for interpreting and understanding their own cultures from an external perspective. I mentor research on Air Force internal and organizational culture, so that the larger organization can benefit from the social scientific lens. In this way, my courses are designed with topics and perspectives quite similar to what I taught in nonmilitary academia, but deeper and more applied, because my students are experienced adults who have traveled the world and have worked closely with individuals in a wide variety of cultures.

## Know that Your Students Are Experienced, Professional Adult Learners

Like adult learners in other professional education settings, military officers are experienced, skilled at their jobs, educated, and often extremely well-traveled. Being an officer in the military does not preclude your students from sharing many of your values. Then again, they might see the world through a lens very different to yours—their worldview based on their experiences and assumptions. They might not agree with you, but they are likely as smart as you are. This is much like teaching in a standard academic setting, where every adult learner brings their own insights, opinions, experiences, needs, and perspectives to class with them. Additionally, recognize that many of the difficulties of teaching in a civilian setting will be mitigated by the fact that military organizational culture values professionalism and focused effort.

## Know Where You Are Positioned

You are both insider and outsider, and that is a valuable role to play. First, think of yourself as an insider: learn how to adapt without having to adopt. You can work effectively within the organizational value system and culture of the Service that employs you for without needing to wholesale adopt the local values. I also offer this advice to military personnel in terms of understanding other cultures' values and beliefs. They are not obligated to adopt the local culture's value system, as that would run counter to how most people operate in the world. Rather, learn to understand those values, recognize where yours might coincide or conflict, and be prepared to bridge a few gaps. Why is this relevant to military culture center faculty?

   I have seen civilian academics be completely tone-deaf to their military employer's organizational culture. If you cannot successfully adapt to the organizational culture of the military branch that employs you, then you will be less than useful in multiple ways. The U.S. Air Force, for example, is a unique cultural setting with its own beliefs, values, symbols, rituals, naming practices, jargon, and mythology. As anthropologists know, respect for another culture will move you toward shared understanding. Show—or rather, feel—that respect by learning about the cultural context you are working within. I was motivated by a sense of excitement at the Air Force's unique culture to explore the Service's slang using a sociolinguistics lens, examining the links between language and culture. I began writing a series of blog entries entitled "Speaking Air Force-fully" in which I apply this perspective to parsing out the origins, meanings, and unique cultural characteristics of Air Force terms. Often, as with any other cultural fact, these terms have been so inculcated into Air Force culture that they are often used and rarely defined. Terms like *spun up*, *recage*, and *reblue* have uniquely Air Force cultural meanings that are often opposite to what the terms appear to mean.[3] It is a simple and grounded tool for me to pull aside the curtain of organizational culture and make accessible to nonanthropologists the ways that the Air Force is a unique cultural setting.

Conversely, cultural misunderstandings happen on the military side as well. I have seen military personnel who do not recognize that what they see as presumptuous or divisive values held by civilian academics are actually fundamental cultural characteristics of the academic world. For example, some of the intellectual diversity provided by civilian academics include the perspective that faculty are organizational co-owners of the academic venture, rather than simple employees. These are fundamental academic cultural values with an emphasis on horizontal power and democratic decision making. While these values bump up against military values—sometimes gently, sometimes titanically—they are not wrong but simply different. These differences point to meaningful perspective

---

[3] *Spun up* and *recage* refer to aspects of using a gyroscope when flying a plane. *Spun up* means to get prepared (start the gyroscope spinning prior to flight) and *recage* is to reset the gyroscope back to its baseline position. *Reblue* is the process of culturally and psychologically reinstating the Air Force identity of airmen who have been in Joint assignments.

checks that can help leadership in PME to better understand their civilian employees and make civilians feel more like a part of the larger organization.

Additionally, an academic will have far more impact if they see their role at an MCC as being an internal cultural advisor. Who are you more likely to take advice and input from—a colleague with whom you have a respectful long-standing working relationship or someone outside the organization with an adversarial viewpoint that sees the military as inherently "wrong"? Working within the system and building respectful relationships with military colleagues leads to effective communication and more impact on your part.

Now, think of yourself as an outsider. Outside is a good place, maybe even the right place, for an anthropologist to be. Being an outsider is a fundamental aspect of doing anthropological work, and any trained anthropologist should be used to this position. We must stand with one foot in and one foot outside of any group that we are attempting to study. There is a reason that we are called "professional strangers."[4] That outsider viewpoint is of tremendous value to the military as are the fundamental skills of the social sciences. We bring expertise in critical observation of human behavior, values, and beliefs; an ability to set aside our cultural assumptions and a tool kit for scholarly analysis of organizations and groups of people. We offer a lens to the military that benefits planning operations, partnership building, and interpreting other cultures.

You benefit the organization by not aligning with existing doctrinal viewpoints. There is strength in diversity of opinions. Do not apologize for the fact that your personal or disciplinary viewpoint does not align with academic colleagues or military doctrine; your value is in offering a different, sometimes opposing, perspective.

---

[4] Michael H. Agar, *The Professional Stranger: An Informal Introduction to Ethnography*, 2d ed. (San Diego, CA: Academic Press, 1996).

## Know How to Be Cognitively and Disciplinarily Flexible

MCCs were typically interdisciplinary. Diversity of disciplinary knowledge and viewpoint enriches our dialogues with our peers and what we offer our students. Learn the key theories and concepts in your peers' disciplines and how those theories intersect, overlap, or conflict with those of your own field. Disciplinary differences play out like cultural differences, so learn to adapt to speaking the theoretical languages of your peer faculty. This is a beneficial skill for any academic working in any type of academic setting.

I learned—and relearned—a significant difference in disciplinary perspective one time in a sociology graduate course and again, years later, in a meeting with military and civilian political scientists. Both times, I heard cultural patterns, activities, or beliefs being referred to as "problems." The problem of X, solving the problem of Y. That is not how I was trained as an anthropologist. What sociology and political science were labeling as problematic were, to my anthropological eye, simply cultural facts. Not problems, just facts. Did the people in the community being discussed see these cultural facts as problems? I had no evidence of that. Did we academics see it as a problem? Apparently, some did, but I did not. It was important, though, for me to understand that there was a fundamental difference of viewpoint between my anthropological self and my peers. I would be better able to translate between our disciplines by recognizing the differences in our lenses.

Also, be prepared for swift action and big changes. MCCs operated at a faster pace than the world of standard academia and required faculty to be flexible and adaptable. New projects were constantly entering the pipeline, with quicker turnaround times than is usual in academia. MCCs are also more orderly and less democratic than typical university settings. This does not mean that your input is ignored; rather, recognize that you will not necessarily have the level of input that you would have at the departmental level in standard academia. Conversely, the setting will be dynamic, the deadlines sooner, the topics highly varied, and it will never feel like the same job twice. The variety, which requires that you learn about new topics and keep your skills sharp, works well to keep faculty motivated and engaged.

While it is disheartening to see the downward swing of the pendulum with the U.S. military defunding and closing up shop on military culture centers, there is an opportunity here as well. We can start looking inward. We can keep applying anthropological analysis to the organization, find ways to help support and offer insight. We can prepare for the time when MCCs are again seen as a necessity. For individuals who want to see their discipline applied in the real world, teaching at a military culture center has a powerful pull. Choosing to do so risks placing yourself at odds with your disciplinary peers, but it also offers the possibility of having a significant, real-world, positive impact on the communities encountered by the U.S. military. None of us will ever know the extent to which our teaching has had an impact on our students' lives or in the lives of the people they interact with. We all operate on a faith that there is value in what we do. If even one less human life is lost because of something I taught, my career at an MCC will have had value.

# CHAPTER FIVE

# The Company I Kept
## Twenty Years at the Naval Postgraduate School

*by Anna Simons, PhD*

## Introduction

It is hard to know which beginning to lead with, so let me start at the end: I recently retired from 20 years of teaching anthropology at the Naval Postgraduate School (NPS) in Monterey, California. NPS was the first professional military education (PME) institution in the United States (and, as far as I know, the world) to hire an anthropologist to teach anthropology full-time to military members. Ironically, I spent several years in the early 1990s lobbying the U.S. Military Academy at West Point, New York, to hire an anthropologist, but could never get the academy to pull the trigger. I also tried the same tactic with the U.S. Army John F. Kennedy Special Warfare Center and School at Fort Bragg, North Carolina. Meanwhile, I had never heard of NPS prior to seeing a small ad in the *Chronicle of Higher Education* in 1998. The ad was for a position in the Special Operations Academic Group, otherwise known as the Special Operations/Low Intensity Conflict (SO/LIC) curriculum. The group was not even large enough to constitute a department at the time, though we eventually became the Defense Analysis (DA) Department.

The SO/LIC program was not looking for an anthropologist in 1998. Cofounders of the program had no idea someone like me existed. Instead, the ad I responded to had been written with a particular individual in mind, someone who was hired at the same time I was offered a visiting position.

Timing being everything, the ad appeared within months of my earning tenure at the University of California, Los Angeles (UCLA). Earning tenure had been politically tricky given the fact that I studied green berets, which is how one of my departmental tenure committee reports described U.S. Army Special Forces, lower case letters and all. Consequently, I was not sure I was ready to throw away what I had worked so hard to earn, especially since UCLA was a top 10, four-field anthropology department, a rarity even in the late 1990s. Was I really ready to give it up for a job at a place no one I knew had ever heard of? No. So I asked my department for a two-year leave of absence, although within the first several months at NPS I knew that I would likely stay.

Hands down, the best part of teaching at UCLA was its undergraduates. My standard line at the time was that while Harvard University—my alma mater—prided itself on diversity, its diversity was manufactured, with the admissions office applying its own predetermined metrics like: we will take one from Wyoming, three from Alabama, six of this color, eight from that background. In contrast, UCLA's diversity was totally organic. Whenever I taught about the Vietnam War, for instance, I could almost always count on having in class some kind of cross section of Vietnam War veterans, sons and daughters of Vietnam War protestors, and students who were Vietnamese- or Laotian-American.

During the course of my six years in Los Angeles, I taught a wide variety of undergraduate and graduate courses. One smart thing UCLA's department did was to *not* allow, never mind make, junior faculty teach the big introduction to anthropology classes. This way we were not overwhelmed at the outset. Nor did we have to try to manage teaching assistants.

Instead, as junior faculty, we taught mostly upper-level electives and graduate seminars, which meant we could introduce new courses into the curriculum. Among those I introduced were two on the anthropology of warfare and conflict. In addition, I taught undergraduate courses about Africa, pastoral nomads, and anthropological methods, and graduate seminars on topics like the social science triumvirate of Émile Durkheim, Karl Marx, and Max Weber. I also devised a seminar on cross-cultural miscommunication for UCLA's Honors Collegium. In fact, had I stayed at UCLA, I would have been one of the few faculty members to teach two seminars in the collegium. The draw of the collegium was that it attracted smart students who clearly liked to read and think, since they too had to apply for admission to the program. Beyond being selective, these seminars were wonderfully small.

I mention all of this to set the stage for what I encountered at NPS, where we offered an 18-month terminal master of science degree in an interdisciplinary field that existed nowhere else—defense analysis.

But to further set the scene, I also need to briefly sketch several other beginnings.

## Shaggy Dog Beginnings
*The Context beneath the Context*

Beginning number two: I rarely enjoyed school. I escaped high school half a year early and completed college in three years. Graduate school never entered my mind. My ambition was to write and to travel. After relatively short stints on a newspaper, writing speeches for President James "Jimmy" Carter during his last year in office, and trying to do the same for the governor of Arizona, I finally became a vagabond. I spent three and a half years working and traveling abroad. The better part of two of those years was spent trekking north to south and then south to north overland in Africa. This is what eventually got me to graduate school.

Beginning number three: I grew up across the Potomac River from Washington, DC, in Alexandria, Virginia, back when "Alexandria" meant nothing to anyone outside of Northern Virginia. Even so, our neighborhood was full of retired and active duty military officers. Friends' fathers deployed on a fairly regular basis, not that I understood what that meant at the time. Two memories stuck with me. First, the Army Navy Country Club had the biggest, nicest pools in the area, which was important in Washington, DC, during the un-air-conditioned summer. Second, I was always made to wait *outside* of the post exchange (PX) and the commissary on the country club grounds whenever the friend who took me to the pool with her went shopping with her mother. The fact that I was not allowed inside (because I did not have an ID card) made the military seem both gloriously mysterious and alluringly exclusive.

As for my first extensive encounters with soldiers, these took place outside the United States in Israel and then throughout Africa. Often in Africa, this was because soldiers and officers were deployed far from home and talking to two young women—a 20-something American, me, and a 20-something Australian, my travel buddy—offered welcome distraction, though not infrequently we also got stuck at checkpoints and talking to soldiers was our way of ingratiating ourselves so that nothing bad happened to us.

But overall, encounters with military forces provided little more than background noise to what really consumed me by the time I entered graduate school: What accounted for such profound differences between the West and the rest?

Beginning #4: through a series of accidents, I ended up in Cambridge, Massachusetts, once I was back in the United States, where I made an appointment with the then-chair of Harvard's African studies program. I wanted to ask her where she would recommend that I go for a master's degree in African studies. My thinking was that maybe *this* would offer me the credentials I needed to publish the screeds about foreign aid that I intended to write. Her response was not what I expected. She wondered whether I would consider continuing in anthropology for a PhD. I did not tell her that the only reason I had majored in biological anthropology as an undergraduate was because biology required too much work thanks to the many premed students in biology classes. The other thought bubble that I kept to myself had to do with studying and theorizing about people as if they were specimens, which held zero appeal. So, I very politely told her I would think about it.

I went home that night and consulted a family friend, who was a prominent political scientist: What about political science? He told me that no graduate school in any discipline would grant me admission for the fall at such a late date; I would have to wait another year before applying. So, that decided it. Impatient youth that I still was, I defaulted to anthropology.

Beginning #5: thanks to my travels, I knew exactly where I wanted to return for fieldwork—East Africa. More specifically, northern Kenya. Anywhere in the Sahel would have been fine, but we had gotten way off the beaten path in northern Kenya and I knew I liked the desert, I knew camel nomads were understudied, and I thought if I focused on them that would help me expose a lot of misguided development aid.

But like all plans, this one went awry in almost every conceivable way. I did succeed in getting back to northern Kenya during my second summer in graduate school. The aim was to line up my fieldwork site and genuflect to all of the right people for all of the necessary research permissions. By the time I had everything in order and was back in Kenya a year later (1988) to head up to Kenya's remote northern reaches to begin classic live-with-nomads fieldwork, the Executive Office of the President in Nairobi decided to deny permission to anyone seeking to do research in northern Kenya that year. I think there were a grand total of three of us at the time.

Fortunately, the news did not come as a total shock; I had been warned that I might have difficulties and had been advised to have a backup plan before I left the United States.

And so, I had a visa in my passport for Somalia. I had managed to affiliate myself with a World Bank project that concentrated on development in the Central Rangelands. Not only did Somalia boast the world's largest camel herds but more than half of the population was said to be nomadic.

Of course, there were just a few minor challenges associated with switching from Kenya to Somalia—like the language. I had not studied Somali. Also, I had never set foot in Somalia previously. But, longest story short, it also became impossible to live with camel nomads. I arrived in late 1988. By July 1989, the civil war that was tearing up the north spilled south. Unrest confined me to the capital, Mogadishu. Consequently, my research focus had to shift. I was already paying attention to all of the ways in which expatriates perceived, or misperceived, Somalis. I also had a sad but sobering front-row seat for how dissolution was impacting the Somalis I knew.

Beginning #6: there were not many expats in Mogadishu in the late 1980s. Among them were four members of a U.S. Army Special Forces (SF) Mobile Training Team (MTT): three noncommissioned officers (NCOs) and a captain. I spent a lot of time with them; they were my introduction to the U.S. military. The team was in Somalia as part of a multiyear train the trainer effort; by 1988, the Green Berets' chief job was to help oversee the Somali trainers. But, of course, their oversight was not exactly going according to plan either since Somalia was falling apart, which only added to the team's frustrations.

Because I was already paying attention to expat frustrations, it was not long before I tried to explain to my new SF friends *why* Somalis were behaving in ways that did not make sense to them and thereby aggravated them. I figured that maybe I could help allay their frustrations. But, as I quickly discovered, I was way too late; their Somali counterparts had already lied to them so frequently that nothing I said was going to change their minds about the character of the people they were in Somalia to work with.

This then prompted me to write my first letter to a general

officer. In the end, the family friend who first suggested that I write this particular general thought better of forwarding my letter, which was probably just as well. But here is some of what I wrote to the head of U.S. Special Operations Command (no less) in February 1990:

> [I]f future teams could be properly armed with the right kind of ethnological information in advance, they might be more likely to find themselves in an inherently frustrating situation without feeling quite so frustrated. I think some of what anthropologists have learned could help SF in Africa, by providing the . . . nuts and bolts of how particular African societies work. Political and military briefings may not be enough. They may not sufficiently prepare a team for an alien culture, no matter how modern or much like ours the host country and its military may seem on the surface. Each country in Africa is unique; even regions within countries can be radically different from one another. Also, Islamic countries in Africa seem to present special problems for Americans, many of whom have deep-seated views (whether admitted or not) about blacks, second only to their feelings about the Muslim religion.
>
> I think an anthropologist could offer SF teams a head start before they ever get to the field. Briefings could serve to warn team members about what they will encounter that they can't expect to understand without first thinking in terms of the dynamics of village-level social organization; what they will encounter that won't make sense, or is "not right" according to American standards, but what can be made sense of using local standards (so that team members at least have a better handle on what

constitutes the local mentality); what they can expect people to want from them, and how subtly or blatantly they should expect to be manipulated; and how they can best handle and/or deflect that manipulation.

A somewhat arrogant letter!

In the letter, though, I also asked General James J. Lindsay whether I could study Special Forces in order to help debunk Green Berets' image as a bunch of Rambos. *That,* at least, I later got to do.[1] Meanwhile, fast forward to the 1998 Naval Postgraduate School ad in the *Chronicle of Higher Education*—the prospect of finally being able to teach Special Operations Forces (SOF) officers seemed too good to be true.

## From UCLA to NPS

I took several lessons up the California coast to NPS with me about what seemed to work best with students:
1. Always assign reading that students will want to do—readable, relatable books. And use books rather than articles; they stick with students better. For better or worse, this also means books written by journalists, the best of which are much more accessible and informative than books written by contemporary anthropologists.
2. If tests are required, make them multiple choice and matching. If the point is to test whether students have done the reading and/or attended lectures, then why make them think and synthesize under time pressure. Written test es-

---

[1] In the interim, I was also able to go to Fort Drum, NY, to take a stab at soldier-Somali relations for a project sponsored by the Walter Reed Army Institute of Research, thanks to Dr. David Marlowe, a Harvard-trained anthropologist who had done his fieldwork in Somalia as well.

says are almost always too painful to decipher. Instead, assign thought papers.
3. Thought papers should be no longer than two to three pages, double-spaced. Anything longer than that and students have too much time to bullshit. Anything shorter and they will not put sufficient thought into what they turn in. The most stimulating questions to ask are provocative questions to which there are no correct answers.

I also took all of my course material, obviously. I knew I would have to modify a good bit of it. For instance, at least one-third of the Anthropology of Warfare and Conflict course at UCLA had been devoted to talking about the U.S. military. I also used to invite one of the Reserve Officers' Training Corps (ROTC) instructors to come to class in his green Class A Army uniform so that a retired soldier could then "read" his uniform for the students. Needless to say, this activity was totally unnecessary at NPS.

## The DA Department

The first course I taught at NPS was the Anthropology of Conflict. The following quarter I taught Low-Intensity Conflict: Africa. Most students at the time were senior O3s.[2] Several had worked or traveled in Africa. Almost all of them had deployed somewhere.

Our students hailed from the various Special Operations

---

[2] O3 designates a captain in the Army, Air Force, and Marine Corps and a lieutenant in the Navy. O4s are majors in the Army, Air Force, and Marine Corps, and lieutenant commanders in the Navy. One reason people use the shorthand of O3 or O4 is to avoid confusion in mixed Service environments since the title captain refers to an O3 in every Service but the Navy, where a Navy captain is three ranks higher than an Army, Air Force, or Marine Corps captain. See "U.S. Military Rank Insignia," Department of Defense, accessed 12 February 2021.

tribes, which means we had officers from Army Special Forces, civil affairs, psychological operations, the 160th Special Operations Aviation Regiment (Airborne), and the 75th Ranger Regiment. We also taught Air Force Special Operations pilots and navigators, as well as officers from U.S. Naval Special Warfare Command, most of whom were Navy Sea, Air and Land Forces, or SEALs. We received a sprinkling of regular Air Force pilots and Navy surface warfare officers as well and served as a test bed for the Navy's Seaman to Admiral (STA) program.[3] Because we had five SEALs who were slated to earn their bachelor's and master's degrees during a three-year period, I got to teach them as many undergraduate-level courses as I could invent; in addition, they took the same classes everyone else did. The first of the group recently made it to admiral; he took approximately nine classes with me, so many that we used to joke at the time that he was majoring in anthropology.

We received cohorts of students twice each year and, in 1998, we consisted of four full-time faculty. Because we operated year-round on the quarter system, we each taught all of the students continuously, which made it easy to build on what we knew we had previously conveyed. Classes were small enough to be run like seminars, though the other way in which we were able to work intensively with students came through advising theses.

By the time I retired, I had advised upward of 135 theses as principal advisor, considerably more than anyone else in the department. Because the vast majority of these theses were unclassified and would reside in the public domain forever, I felt it critical to ensure they were as well-argued and well-written as possible. This goal turned out to be a labor of love for four reasons, all four of which shed light on the uniqueness of our program.

---

[3] For more on the STA program, see "STA-21: Seaman to Admiral Program," Naval Service Training Command, accessed 12 February 2021.

First, unlike a normal graduate program, we had no say over who was admitted to ours. Our job was to teach whomever we were sent. Those sent to us were often command-track, rising stars. But they were not necessarily what some academics would consider to be typical students. Second, graduate school represented a do-over for many of our officers. Most were grateful for a second chance to learn and think in a semistructured setting, and they usually freely admitted that they had not necessarily applied themselves as undergrads. Some, of course, still resisted applying themselves. But, with rare exceptions, even those officers who were most enthusiastic about school seldom retained normal college-level writing skills. Third, everything that was true of our American officers was also true of our international officers. When international officers from Eastern Europe, Asia, Africa, and Latin America began attending the program in 2003 they represented both a gift and a complication. Some countries consistently sent their best and brightest; these individuals added tremendous breadth and depth to discussions. In other cases, individuals came to California thanks to family and political connections, clearly. Among the latter were several who did not merit the degrees they were awarded, at least not scholastically speaking. However, here too, larger equities were at stake, which brings me to the fourth way in which we differed from a normal research university: all of our students came to us after time spent in the real world and all were heading straight back out into an operational environment. We were cognizant of this before 9/11. But after the 11 September attacks, there was no escaping what our students, including our international students, would be doing: they served at the tip of the spear in the Global War on Terrorism (GWOT).

Given where our students ended up and the gravity of their roles, one might wonder what could be more important than exposing them to concepts that might help them better analyze adversaries, allies, situations, and cross-cultural encounters. Here, too, is where having international officers in classes proved to be both a gift and a complication—a gift because they helped me shed light on important cross-cultural misconceptions, but a complication because we could not always discuss everything with equal frankness given their understandable sensitivities.

To describe other wrinkles that impacted what and how I taught culture, I should also say something about other changes over time:

1. Our cohort numbers and class sizes grew. This made it impossible to run everything as a seminar. However, I also learned that not all subjects lent themselves to discussion unless I could be sure that everyone had done all of the reading prior to class, which, again, was an impetus to only assign reading I thought students would enjoy. I became good at figuring out what kind of reading this was, but I still ended up occasionally having to jettison books students told me they could not get through because they were "too flowery" (a.k.a. evocative or wordy), along with reading that was "too annoying" (a.k.a. too reflective of someone else's *contemporary* military experience).

2. For instance, the most popular course I taught—on military advising—could only be taught in small sections; it *had* to be run as a seminar. I first offered this class in 1999 as soon as I realized that no forum existed for the study of advising even though advising represented an essential SOF mission. From the beginning, students preferred historical first-person accounts to anything contemporary. I structured the readings more or less chronologically so that we reviewed the history of advisory efforts, at the same time each highlighted a certain set of issues. I did end up retiring a few books over the

years, but anyone who took the class in 2019 would have read at least some of the same books as those who took it in 1999. This did not just help turn class into fieldwork for me—in terms of how consistently or differently each cohort responded to the same kinds of questions and dilemmas over time—but it also meant that I could invite back former students who had advised or had commanded advisors since taking the class themselves. It was always rewarding to have a Special Operations Task Force (SOTF), Combined Joint Special Operations Task Force (CJSOTF), or Special Forces Group commander come back and be reflective about their experiences.[4]

3. In the late 1990s and early 2000s, our students were O3s, and the preponderance came from the Army. As captains, our Army officers generally found out partway through the program whether they had been selected for resident Army Command and General Staff College and thereby could consider themselves in the top half of their year group with better than average career prospects. You could see all of them begin to recalculate accordingly, but none became especially cynical. Then, for much of the GWOT, virtually all of our students were O4s, and as requirements for intermediate-level education (ILE) changed, resident ILE was no longer a discriminator, which meant that, as majors, our Army students never knew exactly

---

[4] A SOTF, or Special Operations Task Force, is typically overseen by an O5 (Army lieutenant colonel or Navy commander). A CJSOTF, or Combined Joint Special Operations Task Force, is typically overseen by an O6 (Army colonel or Navy captain).

where they stood vis-à-vis one another and consequently they expended considerable energy in extracurricular networking and politicking. This became one unfortunate source of cynicism, though far more pernicious was what was transpiring—or not transpiring—in Afghanistan and Iraq.[5] By 2018, student cynicism was so palpable and so extensive regarding the wars, senior leaders, and policy making in general that there were very few topics we could not discuss. This marked a sea change, especially considering that right after 9/11 it had been impossible to even vaguely suggest that the 9/11 hijackers were anything but cowards. By 2018, it was totally acceptable for me to refer to at least some jihadis as "true believers."

4. However, whereas analyzing and critiquing U.S. foreign policy and national security strategy became easier over time, referring to domestic American politics grew harder. Again, for at least the first several years after 9/11, students did not want to hear anything critical said about President George W. Bush or his policies. But then, with the 2008 election, politics became a minefield in the classroom. Unless students already knew where each other's heads were, they said very little that might indicate they leaned one way or another along the conservative-liberal spectrum. As it happens, the faculty also became more politically riven, though our deepest differences had more to do with the prosecution of the wars in Iraq and Afghanistan and where we saw Washington erring, some but

---

[5] Anna Simons, "Cynicism: A Brief Look at a Troubling Topic," *Small Wars Journal*, 16 February 2021.

not all of which was colored by the disciplinary lenses through which we analyzed both.

5. We were a very unusual interdisciplinary department, less because we seldom agreed with one another about how best to conduct, or even study and analyze, counterinsurgency, counterterrorism, and irregular warfare—our raison d'être—than because we never took our "shoulda-woulda-coulda" disagreements out on the students. Instead, we exposed our officers to wildly divergent and often contradictory points of view. Students benefited tremendously from this, though it did occasionally create difficulties when first and second readers on a thesis disagreed about a student's approach. Even so, the best among us routinely deferred to whatever approach the *student* wanted to take since this was their thesis. I should add that there was an overall gender/prior service/disciplinary bias that consistently ran through the department: while male faculty acknowledged that "culture" was important, they never considered it quite as important as "strategy" or whatever subject they happened to teach.

Over time, two additional changes occurred in *who* we taught: warrant officers and noncommissioned officers (or senior enlisted) entered the program in small but still significant numbers, and just before I retired, SF officers no longer dominated in quite the way they had previously; they were also O3s again rather than O4s.

One final wrinkle I should mention has to do with the small size of the SOF community. Elsewhere, I have quipped that "reputational vetting" is a SOF operators' favorite pastime. Not surprisingly, because we taught so many officers, we too earned reputations. In fact, it would be easy to trace the lineages of students we taught based on which of their elders we had in classes and who steered their protégés our way. I was always lucky. I benefited from timing (my longevity), the subject matter I taught—culture—and the readings I assigned.

## What Teaching Taught Me

In truth, though, I never did really teach about culture. I actually forbade students from using the word "culture" in classes—the only word I disallowed. I did so because I wanted them to have to work through why people X might do such foreign-seeming things. I did not want them to default to using "culture" as a black box term that explains everything and nothing at the same time.

My job, in my view, was to help our students learn how to unpack others themselves. My reasoning was that our students were all adults. If they did not want to engage with the subject matter, I was not going to be able to make them. So, there was no point in using tests. I assigned books, we watched documentaries and movies, and they had to write me short thought papers. All of this usually came as a shock to them, and they initially distrusted me when I said I was not interested in having them repeat back to me anything I said. Instead, I was interested in what they thought, and I did my best to provoke them to think differently and make me think differently too.

I also felt it was a disservice to spoon-feed our students prepackaged anything. I knew that they generally craved the bottom line up front and had an outline- or PowerPoint-driven need for frameworks and takeaways. At one point fairly early on, I remember being asked if I could just give them the "3 x 5 card" summary of whatever I was trying to convey. Inwardly, this made me cringe. Here were the military's preeminent practitioners of the unconventional, and they were so used to linear approaches and bullet points that they not only did not recognize how conditioned they already were, but I had to figure out how to get them to *want* to relax. Fortunately, I had some credibility thanks to time spent in Somalia and elsewhere in Africa, so I could generally get them to give me the initial benefit of the doubt. I did my best to inoculate them against frustration by explaining that everything *would* connect by the end of the quarter, and I learned that it helped if I did occasionally provide them with a framework.

For instance, I updated and turned the classic ethnographic approach of beginning with the local ecology as the underlay for people's way of life into something they could carry away with them and apply more broadly. I walked them through how to play with concepts like Big Man and Chief.[6] Was, for example, the president of the United States a Big Man or a Chief? What about an O3? I similarly stretched terms like *acephalous* and H. H. Turney-High's *military horizon* to see how far we could push these ideas and whether they could help us reframe conventional thinking.[7]

I tried to remember to write on the board during my first meeting with new students a trio of aphorisms:
- Everything connects, which I treated as an anthropological truism.
- It all depends, which I told students would be the correct answer to almost anything I or anyone else would ask them.
- You just never know, which was a talismanic reminder that no matter how trivial or esoteric something might seem it could still prove useful one day.

I never assigned theory. I did not see the point since our students were not being educated or trained to become anthropologists. Instead, I walked them through theoretical ap-

---

[6] Lamont Lindstrom, " 'Big Man': A Short Terminological History," *American Anthropologist* 83, no. 4 (1981): 900–5.
[7] Harry Holbert Turney-High, *Primitive War: Its Practices and Concepts* (Columbia: University of South Carolina Press, 1949; repr., 1991).

proaches whenever I thought these were relevant, and I did so too as a way to subliminally remind them that the study of others is not always as straightforward as they often assumed. To this end, I also exposed them to at least some anthropological classics, such as Ruth Benedict's *The Chrysanthemum and the Sword*, Lincoln Keiser's *Friend by Day, Enemy by Night*, and E. E. Evans-Pritchard's description of fieldwork among the Azande.[8] One reason I assigned texts like these was to experientially teach students how much invaluable information can still be gleaned from: a) books, b) old books, and c) accessibly written old books, despite how dated they might seem. This became all the more pressing once laptops appeared and everyone turned to Google during discussions so that they could one-up one another and me with information that they often were not knowledgeable enough to properly vet. With laptops open, too, many students also could not help but engage in a weird form of competitive hyperlink hopscotch. Eventually, I banned laptops and tablets in seminars.

While it was critical that our students learn how to better vet sources, it was clear that it made no sense to expect them to remember what the differences were between Japanese and Kohistani (Pakistan) notions of honor, for example. One faculty trap I did my best to avoid was to assume students would be able to retain the same information I did. Just because I drew on the same material repeatedly, and it was germane to me, did not mean that the students absorbed it in the way I intended. I also knew students mentally dumped information at a prodigious rate. I did my best to head this off by never giving them exams and by telling them up front that there was

---

[8] Ruth Benedict, *The Chrysanthemum and the Sword: Patterns of Japanese Culture* (Rutland, VT: Tuttle Publishing, 1946; repr., Boston, MA: Houghton Mifflin, 1989); Lincoln Keiser, *Friend by Day, Enemy by Night: Organized Vengeance in a Kohistani Community* (Fort Worth, TX: Holt, Rinehart and Winston, 1991); and E. E. Evans-Pritchard, *Witchcraft, Oracles, and Magic among the Azande* (Oxford, UK: Clarendon Press, 1976).

no reason for them to commit any ethnographic information to memory. Instead, we were going to use ethnography during our discussions for compare and contrast purposes.

Concepts, concepts, concepts—that is what I sought to push. And I pushed questions. In fact, along with hoping that students would retain an appreciation for certain concepts, as well as an appreciation for context—and an appreciation for the significance of context *as* a concept—I wanted them to walk away with at least one overarching, scalable, stretchable question. Ideally, this question would act as a mnemonic device and would remind them about what they should want to learn, no matter who they were interacting with.

For instance, the takeaway question from Anthropology of Conflict was: What makes an X an *X*? This grew out of our discussions during the course of the quarter about identity, values, and people's priorities. For example, we comparatively and recursively tackled what made Japanese *Japanese* (ca. World War II), Germans *German*, Americans *American*, etc., along with what makes radicals *radical*, moderates *moderate*, and SEALs *SEALs*, or SF *SF*.

The question I distilled out of the cases we examined in Low Intensity Conflict: Africa was: Who is where vis-á-vis whom, and what? One of the things I hammered hard in that course was the significance of both literal and figurative positioning (e.g., in terms of resources, geography, demography, socioeconomics, etc.) as well as timing. I sought to drive home the idea that no country or conflict should ever be considered in isolation. For instance, it is impossible to understand the 1994 genocide in Rwanda or its aftermath without also studying events and dynamics in Burundi, Zaire/Congo, and Belgium prior to and after independence, Cold War politics, and the list goes on. Or as one of our Pakistani officers pointed out every time he was subjected to the term *Af-Pak*, which clearly grated on him: What about India? China? Russia? The -stans? The United States? And untold corporate players?

The question I devised for Political Anthropology: Methods of Social Control was the corollary to who is where vis-á-vis whom, and what? This corollary question was: Who can do what to whom, using what? Meanwhile, the more orthogonal our subject matter, the stickier each of these questions became, so that reading books about bananas, Henrietta Lacks, and Robert Mugabe, or seeing movies about Rumspringa and Wounded Knee still resonate with at least some of my former students.[9]

## Research and Other Opportunities

Whenever I visited former students in Iraq, Afghanistan, and elsewhere, I was mindful of the value of questions from a wholly different angle. When I was at a SOTF, CJSOTF, or on a firebase, I was still a professor, but *I* was now the one out of my element in our graduates' realm. I knew better than to offer my two cents. But I still could not stop myself from asking questions. Sometimes I asked questions to which I knew I did not know the answer. Sometimes I asked questions to which I thought I knew the answer. And sometimes I posed questions that I was pretty sure no one else had yet asked a commander. Because my asking questions was expected, I learned over time that this was also the most useful way for me to be suggestive. I saved most of my critiques and observations for scraps of paper in my pocket or for classes, which is one of the reasons graduates in command positions invited me into the field; they wanted to make sure I stayed up-to-date on

---

[9] Dan Koeppel, *Banana: The Fate of the Fruit that Changed the World* (New York: Plume, 2008); Rebecca Skloot, *The Immortal Life of Henrietta Lacks* (New York: Gale/Cengage, 2010); Peter Godwin, *The Fear: Robert Mugabe and the Martyrdom of Zimbabwe* (New York: Little, Brown, 2011); a documentary about the Amish and rumspringa as seen in *Devil's Playground*, directed by Lucy Walker (New York: Stick Figure Productions, 2002), 77 min.; and a movie about the 1890 Wounded Knee massacre as in *Bury My Heart at Wounded Knee*, directed by Yves Simoneau (Montecito, CA: Wolf Films and Travelers Rest Films, 2007).

the operational environment students were coming from and would be returning to.

Getting to visit units downrange and seeing former and future students do their thing was one of the great rewards of the job. It also provided the most vivid possible reminder that we were participating in a mutual educational enterprise. Not only did trips to the field grant me a deeper understanding of the challenges teams and staffs faced, but my willingness to visit cinched any number of relationships. These visits also constituted a form of fieldwork that I then fed back into projects for the Office of the Secretary of Defense (OSD) and others.[10]

Here, too, is how NPS proved to be unusual. As faculty at DOD's only research university, we were expected to bring in reimbursable research dollars, which meant finding a sponsor, which meant coming up with relevant and timely projects year after year. The upside to this was that, although filling out the requisite paperwork for travel and research grew increasingly onerous over time (particularly given our Orwellian Institutional Review Board), successful projects led to other successful projects.

My most consistent sponsor was the Office of Net Assessment in OSD (ONA has often been described as the Pentagon's internal think tank). Initially my deliverables were papers. Over time, I began to run sponsored long-term strategy seminars. I would select a cross section of students to join me for one or two quarters on a project of ONA's or my choosing, and the students and I would then brief our results both in Monterey and in the Pentagon.[11] One of ONA's aims was to enable promising mid-career officers to think at the strategic level. One of my aims was to tackle something that no one yet

---

[10] See, for example, Anna Simons, *21st Challenges of Command: A View from the Field* (Carlisle, PA: Strategic Studies Institute, 2017).

[11] Among project topics were: strategic blindside, regional stability, SOF 2030, SOF in China, strategic ambush, and existential fears.

had given sufficient thought, so that it was not just me transmitting predigested ideas to students, but all of us working together as a tiger team.

As often as I could, I combined these seminars with research efforts. In doing so, I borrowed from a different set of NPS experiences. Beginning in the early 2000s, I participated in a number of civil-military relations seminars. Michael Mensch, a retired Army colonel, and the Africa program director of the Center for Civil-Military Relations (CCMR), located at NPS, had spent 15 years living and working all over the continent as a defense attaché—he was the practitioner-facilitator. To help him conduct weeklong workshops on the continent, he always tapped a civilian and usually a female Africanist to travel with him.

This model of a male practitioner paired with a female academic did not just work extraordinarily well but made a deep impression on me because I saw it significantly impact our seminarists. Thus, when I was asked by a deputy assistant secretary of defense to undertake a project on the Horn of Africa, I immediately wrote into the budget travel money so that I could take one of our students as a practitioner-researcher with me. I then did the same for subsequent projects. Indeed, unless I was traveling to visit former students in Iraq or Afghanistan, I always took at least one and sometimes two students abroad with me. They invariably viewed things sufficiently differently from me so that our synergy paid untold dividends, whether we were looking into India's counterinsurgency lessons learned or South Koreans' existential fears. In fact, one such project led one of our graduates to suggest to his command that they sign up everyone in our program who had prior experience in East Africa to work on a yearlong project for that command. This project led to multiple research trips for students, including one that enabled a Tanzanian colonel to take two American majors back to Tanzania with him. Each of these majors was then assigned back to East Africa after NPS, so that the initial project redounded in multiple ways.

In addition to engaging in direct research efforts for OSD and various commands, I also participated in the Regional Security Education Program (RSEP), which was also run out of NPS. Much as CCMR's Africa program represented one of the best uses of tax dollars I had seen in Africa—since CCMR's only aim was to facilitate senior members of a host nation's military, government, and civil society meeting together, often for the first time—RSEP was another win-win-win. In response to the USS *Cole* (DDG 67) bombing in Yemen in 2000, a team of two to four academics rode with every Amphibious Ready Group (ARG) and Carrier Group on their transit across the Atlantic or Pacific.[12] The aim was to provide regional orientation to wherever the group might be headed: the Middle East, the Horn of Africa, or East Asia. Ironically, RSEP lectures were really intended for the ship and the ARG or Carrier Group staffs, but since Navy officers were usually the busiest individuals on board undermanned ships, it was invariably the Marines and naval aviators who attended our talks most often instead.

As with my visits to U.S. bases and theaters overseas, nothing I could learn secondhand, either by reading or by talking to sailors and Marines about shipboard life, would have granted me the same insights as did getting to live aboard ship or sharing a cramped stateroom with women who were decades younger than me. So, yes, while I was an educator on these floats, I was also continually and continuously being educated myself. This was true no matter which component of the military I spent time with—an aspect of teaching in PME that should be considered essential, especially for those of us in the social sciences. Otherwise, how can we gauge what is most pertinent, let alone determine how best to convey what is most relevant to our customers or consumers?

Of course, I never did regard members of the military as either customers or consumers. Instead, they always repre-

---

[12] Or, in my case once, I was a team of one.

sented the thin line in the sand between all of us civilians and harm, so I always considered it self-interested on my part to want them to adeptly handle anything or anyone they might encounter abroad. This is what struck me initially in Somalia in 1989: all four members of the Mobile Training Team there were technically proficient. But the MTT was not especially well prepared—none of the four tried to see either Somalia or themselves through Somali eyes. They also did not have anyone other than their Somali counterparts to assist them with making sense of what they saw.

This is one reason predeployment briefings have always struck me as deficient. Here is where the RSEP program got things a little more right: lectures were offered during a five-, six-, or seven-day period; no one was subjected to a check-the-block session on region and culture during a suite of other predeployment trainings. This also made RSEPs a better approach than SOF's post-9/11 notion of an "Academic Week," when people like me would be given a two-hour slot during which we were expected to condense highlights from already ridiculously condensed quarter-long courses. Whenever I did these sorts of lectures, whether about all of sub-Saharan Africa or *just* about Somalia, I was all too acutely aware of all of the things I was not saying—and was not able to say. I also knew how little of what I relayed was likely to stick.

Thus, if I could wave a magic wand, I would still want the military to do what I first thought it should do after sitting through my first predeployment briefing 29 years ago, when I was a fly on the wall: have a regional "expert" or two meet with the team or group *once* it is deployed. Let everyone recover from jet lag. Let those who are visiting country X for the first time sniff the air and get a sense of the place, and then, during that initial 48- or 72-hour window when first-time visitors are usually most open-minded and keen to *want* to make sense of the strangeness around them, bring in the experts. Experts need to interact with deployed forces just *before* erroneous impressions start to gel, especially since everyone in uniform has been conditioned to think so linearly and in such American-centric ways.

## Conclusion

Of course, I would also be remiss if I did not add the caveat that cross-cultural expertise itself is a misnomer and determining who is an expert is always problematic. As anyone who has been around the military for any length of time knows, even military culture changes. Thus, the best any student of a place, or organization, or set of organizations can hope to do is to develop as much familiarity as is possible and then strive to keep a finger on the pulse. We all develop shortcuts, usually via trusted written or flesh-and-blood sources. The catch is that we need the wherewithal to stay up to date, which takes time. The catch with time for those working in the DOD is that time usually comes out of hide. Or, this certainly has been true since 9/11.

In fact, one of the more serious downsides to NPS's overall academic model, one that became chronic once everything ratcheted up in the wake of 9/11, is that there were always more opportunities than there was time. It was never possible to truly dig in or build on academic work in anything but a short burst followed by short burst manner. Sometimes it was possible for me to circle back to the Horn of Africa or to some other issue or problem. But as a researcher, I was always sponsor-beholden. I did not necessarily work on what *I* thought was most important, though I did my best to lobby for what I thought was most pressing. I also have to say, I was extraordinarily lucky. Mr. Andrew Marshall, the director of the Office of Net Assessment (ONA), for whom I did most of my work, had long been interested in anthropology. In fact, it was Lionel Tiger (an anthropologist of considerable stature) who first brought me to Mr. Marshall's attention; Dr. Tiger had worked with ONA for years. This meant he was also in then-secretary of the Air Force James G. Roche's orbit since Secretary Roche, too, was an ONA denizen. Together, both these men were close to General John P. Jumper. No wonder General Jumper was predisposed to want to infuse more language-and-culture education and awareness into the Air Force when he was its chief of staff.

So, how encouraging was it that I found myself one fall day, in General Jumper's executive dining room with more general officers arrayed around the table than I had ever seen in one place. I had been invited to explain to him and to them what the Air Force might gain from the same sorts of things I was teaching at NPS. Talk about easy: I had never had a problem proselytizing on behalf of what I knew our students found useful. Student enthusiasm alone was a testament to how much value officers found in being able to bring anthropological concepts to bear on the world at large and to their line of work in particular. Nor, thinking back to that day, would I say that the value of anthropology has diminished—at all.

Today, Anthropology of Conflict remains among the Defense Analysis Department classes our flag officer and recent graduates alike cite as having been one of the most valuable they took. It also happens to be one of their favorites while they are taking it. Meanwhile, who is teaching it in my wake? The "everything connects" answer is an anthropology PhD who happened to take one of the very first iterations of the course from me back when he was an undergraduate at UCLA. Unlike me, Siamak Naficy stuck with biological anthropology; and though he did not do fieldwork abroad, he was born in Iran and remains fluent in Farsi. I knew he had an excellent reputation teaching at one of Southern California's best community colleges, so when he followed his future wife up the coast to Santa Cruz, I thought aha, who better to help me teach our students. Proof, after a fashion, that not only does everything connect (and all depend), but you just never know. Except, I did already know Siamak and how talented he is. I also knew that anthropology was now considered a core discipline in the Defense Analysis Department thanks to the foresight of the individuals who decided to hire me, thanks to our students' enthusiasm, and thanks to several colleagues who taught related courses about culture as well as about low intensity conflict in regions other than Africa.

In retrospect, I would still contend that culture was never considered to be *as* important in our department as political science or strategy, probably because it does not lend itself to 2 x 2 tables. But it turns out that anthropology's unformulaic nature made it that much more intriguing to our students, especially since they knew they were going to have to go back out and operate abroad. That reality alone played to Siamak's and my strengths, which were classically anthropological: we got to be a little bit heretical, a little bit irreverent and, thanks to our bottom-up/inside-out stance, we could not help but try to be *disarmingly* provocative too.

CHAPTER SIX

# From Concept to Capability
## Developing Cross-Cultural Competence through U.S. Air Force Education

*by Brian R. Selmeski, PhD*

## An Unexpected Invitation

I arrived at the 2006 Society for Applied Anthropology annual meeting in Vancouver, British Columbia, with only my carry-on bag. At the time, I was a U.S. academic employed at the Royal Military College of Canada, leading an applied research project in Bolivia to help open their army's officer corps to indigenous peoples and women.[1] I was not terribly surprised at my predicament, considering I had flown from La Paz, Bolivia, to Lima, Peru, to Houston, Texas, to San Francisco, California, to Vancouver. I was more startled by the cryptic message that awaited me at check-in: "Join me for dinner and a good cabernet tomorrow? An interesting opportunity has come up." I quickly realized it was neither an invitation to a romantic rendezvous nor a practical joke. Rather, the message was from Dan Henk, an affable retired U.S. Army colonel with a PhD in anthropology who had grown up in Africa

---

[1] Brian R. Selmeski, "Indigenous Integration into the Bolivian and Ecuadorean Armed Forces," in *Cultural Diversity in the Armed Forces*, 1st ed. (London: Routledge Press, 2006), 48–63.

and was on the faculty of the Air War College. I accepted the invitation, hoping it would distract me from my travel woes, but never imagining it would radically alter the course of my career.

I had met Henk for the first time five months earlier in Chicago, Illinois, at the 2005 biennial meeting of the Inter-University Seminar on Armed Forces and Society (IUS). Founded by the eminent military sociologist Morris Janowitz, I found IUS conferences to be an intoxicating mix of international scholars, practitioners, and students—all focused on the military. It was there that one of the coeditors of this volume, Kerry Fosher, and I organized an audacious roundtable on establishing military anthropology as a subdiscipline.[2] Henk had been one of the 20-or-so attendees, but this was the first I had heard from him since.

The "interesting opportunity," he explained when we met, was to help establish the U.S. Air Force Culture and Language Center (AFCLC) at Air University in Montgomery, Alabama. Military operations in Iraq had deteriorated, with the bombing of the al-Askari Mosque the previous month precipitating what some observers were already calling a civil war.[3] The chief of staff of the Air Force (CSAF), after learning about efforts to teach Arabic at Marine Corps University, had provided funds to do the same at Air University. During dinner—and yes, a good bottle of cabernet sauvignon—Henk sketched out his vision: to create an organization that would both teach foreign

---

[2] Although we did not succeed in this endeavor, the roundtable did expand the Military Anthropology Network (Mil_Ant_Net) I had established in 2003, "an on-line community of practice that facilitates free and informed exchange between academics and practitioners on matters related to culture, anthropology and the security sector—all broadly understood." The network was extremely active in its early years, but it has gone dormant more recently.

[3] Robert F. Worth, "Blast Destroys Shrine in Iraq, Setting Off Sectarian Fury," *New York Times*, 22 February 2006.

languages to general purpose airmen and bring anthropology to the military masses. He invited me to hear him present his paper the next day and promised to be in touch.[4] I was hooked—this was precisely the sort of organization I would have gravitated to when I was in uniform and the kind of opportunity I had been told (by a more contemporary doyen of military sociology) would never happen!

## The "It"
*Cross-cultural Competence*

Implementing Henk's vision was easier said than done. The Department of Defense (DOD) already had a robust and effective, if not always efficient, approach to teaching foreign languages. The Defense Language Institute Foreign Language Center (DLIFLC) and its predecessor organization, the Army Language School, have been housed at the Presidio of Monterey, California, since 1946. The institute employs a large faculty of highly qualified instructors, who teach specialized military personnel, from foreign area officers to enlisted cryptologic linguists, more than a dozen languages of strategic importance. They also assess military personnel's mastery of foreign languages through the Defense Language Proficiency Test and Oral Proficiency Interview. These instruments, honed over the decades, are scientifically validated measures of listening, reading, and speaking abilities. In the metrics-obsessed U.S. military, this provided foreign language enthusiasts a distinct advantage in the bureaucratic knife fights over funding. And Arabic was what the CSAF wanted Air University

---

[4] Dan Henk, "An Unparalleled Opportunity: Linking Anthropology, Human Security, and the U.S. Military" (presentation, Society for Applied Anthropology 66th Annual Meeting, Vancouver, BC, 28 March–2 April 2006).

to teach, or in Henk's words, "what he thought he wanted."[5]

Culture, by contrast, was often portrayed as "soft," "abstract," and "squishy," not to mention "tactical." The U.S. Air Force—with its global reach and technological dependence—favored the hard, concrete, and strategic.[6] To compete with foreign language learning, we would have to find ways to measure cultural proficiency. First, though, we would need to determine what aspects of anthropology were militarily relevant. "We know military personnel need it," Henk would say, "but what's the 'it'?"

For the next 18 months, as the inaugural director of AFCLC, Henk generously supported my research.[7] When I returned home that spring—after flying back to Bolivia, the airline losing my luggage again, and wrapping up my work there—I embarked on a massive literature review. As the U.S. military struggled to adopt a counterinsurgency approach in Iraq, culture became a hot topic and professional journals brimmed with experientially informed, a-theoretical proposals. Undeterred, Henk sent me and Kerry Fosher to visit U.S. military educational institutions, their nascent culture centers, and symposia from California to Kansas to Rhode Island.

There was, of course, a small corpus of anthropological studies of the military. I counted the authors of many such works as friends, mentors, or both. There were fewer accounts

---

[5] While Henk and I, both polyglots, agreed that developing deep expertise in a particular culture requires a mastery of the relevant language(s), teaching the culture(s) of Iraq and Afghanistan through Arabic and Pashto struck us as ludicrous. The U.S. military's introductory courses to these languages were 64 weeks long and focused on vocabulary, conjugation, syntax, etc. Professional military education programs lacked the necessary time, the return on this investment was dubious, and the results of initial foreign language learning were highly perishable.

[6] Carl H. Builder, *The Masks of War: American Military Styles in Strategy and Analysis* (Baltimore, MD: Johns Hopkins University Press, 1989).

[7] He also brought one of the editors of this volume, Kerry Fosher, to Air University as visiting faculty, and he hired the other editor, Lauren Mackenzie, several years later. It is no exaggeration to call Dan Henk a godfather of recent U.S. military culture programs.

of anthropologists who sought to teach soldiers, sailors, or airmen our concepts and methods. Most members of the discipline—save for some archaeologists and the old guard at the Human Relations Area Files (HRAF) in New Haven, Connecticut—exhibited only antipathy toward our desire to make anthropology relevant to servicemembers.[8] I found other disciplines trafficking in culture were less averse to the idea, though most would grapple with questions of professional ethics as the ill-named Global War on Terrorism progressed.

Thanks in large part to my experiences at the Canadian Defence Academy, I also broadened the focus of my search, examining efforts in other professions to improve their members' abilities to work effectively across cultural differences. Nursing, social work, and related healing professions were extremely informative but difficult to translate into a military context. Not surprisingly, I found the work of faculty at business schools with international programs to be more practical. These scholars, mostly psychologists and organizational behavioralists, researched and published at the intersection of academia and praxis. Although they collaborated across international boundaries, two countries stood out: the more popular, if superficial, dimensions of culture approach emanated from the Netherlands; and the newer and more sophisticated

---

[8] This occurred despite the efforts of Kerry Fosher and other anthropologists I admired who served on the American Anthropological Association's awkwardly named commission. James Peacock et al., *Commission on the Engagement of Anthropology with the US Security and Intelligence Communities: Final Report, November 4, 2007* (Arlington, VA: American Anthropological Association, 2007). Yet, shortly after they delivered their first report at the association's annual business meeting, I and the other members of a panel addressing how to engage with the armed forces, including two senior scholars long committed to the peace movement, were publicly—if hyperbolically—denounced as "war criminals." Robert A. Rubinstein, "Master Narratives, Retrospective Attribution, and Ritual Pollution in Anthropology's Engagements with the Military," in *Practicing Military Anthropology: Beyond Expectations and Traditional Boundaries*, ed. Robert A. Rubinstein et al. (Sterling, VA: Kumarian Press, 2013).

cultural intelligence approach had its origins in Singapore.[9] Henk made it possible for me to visit and learn from scholars in both these small states, many of whom were collaborating with their militaries.

As the project began to wrap up, I was disappointed we had not found any historical examples of U.S. government culture training programs informed by bona fide scholarship. Then, as I wandered the stacks of a used bookstore in Kingston, Ontario, I spotted a copy of Edward T. Hall's *The Silent Language*. A blurb by eminent anthropologist Margaret Mead caught my eye: "Dr. Hall has probably had more experience than any other American anthropologist in trying to teach cultural difference to people who did not want to learn about them." That sounded remarkably similar to the resistance Henk and I were encountering from some military personnel. Hall's biographical sketch further piqued my interest, particularly that "during the crucial years of the foreign aid program in the 1950s, he was Director of the State Department's Point IV Training Program."[10]

Soon, I tracked down a copy of the manuscript describing the model that guided this training.[11] Not long thereafter, I visited the U.S. Foreign Service Institute, where I found Hall was both venerated as an icon and largely ignored in practice. Then I traveled to the University of Arizona in Tucson to review his papers in the library's Special Collections. Although I

---

[9] Geert Hofstede, *Culture's Consequences: International Differences in Work-Related Values*, 1st ed. (Thousand Oak, CA: Sage Publications, 1980); and P. Christopher Earley and Soon Ang, *Cultural Intelligence: Individual Interactions Across Cultures* (Stanford, CA: Stanford University Press, 2003).

[10] Edward T. Hall, *The Silent Language* (Garden City, NY: Doubleday, 1959). The Point Four Program was a forerunner of the Foreign Operations Administration, the International Cooperation Administration, and eventually the U.S. Agency for International Development.

[11] Edward T. Hall and George L. Trager, *The Analysis of Culture* (Washington, DC: Foreign Service Institute and American Council of Learned Societies, 1953).

never found syllabi or lesson plans, the insights were essential to synthesizing the voluminous materials I had collected. By October 2007, I presented the results as a white paper, which was published by AFCLC the following May.[12]

I concluded that "short-term operationally focused responses to the military's 'cultural needs' emphasize facts over concepts primarily through training," whereas "long-term institutional approaches—such as developing cross-cultural competence—will require a distinct approach."[13] The remaining 24 single-spaced pages aimed to "clarify the concept (what is it that we seek to develop), then craft a framework (the wider set of ideas that inform the objective), and finally establish broad objectives at various professional development levels (a matrix of what military personnel should be, should know, and should do)."[14]

## The Vehicle for Change
*The Quality Enhancement Plan*
Henk was thrilled with the report. I had not resolved the metrics challenge, but I had identified elements of our discipline—and others—that were relevant to the military and offered a conceptual model. Cross-cultural competence, I posited, would permit military personnel to learn about another culture in their own language, using a generalizable approach—"culture-general"—that could be put into prac-

---

[12] Brian R. Selmeski, *Military Cross-Cultural Competence: Core Concepts and Individual Development*, Armed Forces, and Society Occasional Paper Series No. 1 (Kingston, ON: Royal Military College of Canada Centre for Security, 2007).
[13] Selmeski, *Military Cross-Cultural Competence*, 3.
[14] Selmeski, *Military Cross-Cultural Competence*, 2.

tice, and would foster adaptability.[15] It was the intellectual antithesis of the "dos and don'ts" training approach that had flourished recently in the armed forces and the area studies educational approach that took root in academe during the Cold War.[16] He began sending the white paper—and me—to his network across the Department of Defense. Civil servants in the Pentagon were usually too busy to read it. The foreign language set was largely dismissive.[17] Most uniformed personnel found it too academic . . . and long. I assumed my work had fallen flat and that this heralded the end of an unexpected but exciting chapter of my career. I told myself it was time for me to get back to Latin America anyway.

Hence, I was surprised to hear that Air University's commander and vice president for academic affairs had seized on my ideas. In the preceding years, leaders had consolidated the independently accredited units and degree programs, eventually gaining institutional accreditation by the Southern Association of Colleges and Schools Commission on Colleges (SACSCOC) in 2004. Reaccreditation—or "reaffirmation of accreditation," as SACSCOC calls it—normally occurs every 10 years; however, the first cycle is only 5 years. In addition to the usual requirements that address governance, finance, institutional effectiveness, faculty, library, and the like, SACSCOC

---

[15] I defined *cross-cultural competence* (3C) as "the ability to quickly and accurately comprehend, then appropriately and effectively act, in a culturally complex environment to achieve the desired effect." To this, I added two caveats: "1. despite not having an in-depth knowledge of the other culture; and 2. even though fundamental aspects of the other culture may contradict one's own taken-for-granted assumptions/deeply held beliefs." Selmeski, *Military Cross-Cultural Competence*, 12.

[16] The latter can be traced back to Title VI of the 1958 National Defense Education Act, which established Foreign Language and Area Studies Centers (now National Resource Centers) at U.S. universities.

[17] One senior foreign language official eventually agreed to read my white paper, an act she characterized as "the gift of [her] time!" Unsurprisingly, she ignored the findings and recommendations.

requires all institutions undergoing reaffirmation to undertake a Quality Enhancement Plan (QEP): "A carefully designed and focused course of action derived from the institution's existing planning and evaluation processes that addresses a well-defined issue directly related to enhancing specific student learning outcomes and/or student success."[18]

Air University's commander at the time, Lieutenant General Stephen R. Lorenz, was a visionary leader who saw this as an opportunity: The institution could use the chief of staff's funding to implement a QEP on cross-cultural competence. This would both help prepare military students for the challenges of current operations and satisfy accreditation requirements. It would be a win-win-win, General Lorenz insisted. In April 2007, the Council of Deans recommended "Cross-culturally Competent Airmen" as the focus of Air University's 2009–14 QEP, despite their concerns about measuring student learning.[19] Henk was elated and set about recruiting me in earnest. Eventually, he succeeded; I started the Monday after Thanksgiving.

Although my position was funded by the Culture and Language Center, initially I was assigned to the Office of Academic Affairs. My primary task was to transform the white paper into an institutional plan rigorous enough to pass peer review but not too difficult to implement. My implied task was to learn everything possible about Air University, its faculty, staff, and academic programs, to ensure the QEP's success. My secondary task was to help Henk stand-up AFCLC. Any of these activities could have been a full-time job; 2008 was a blur of late nights, frequent travel, incessant meetings, and unremitting PowerPoint slides.

---

[18] *Resource Manual for the Principles of Accreditation: Foundations for Quality Enhancement* (Decatur, GA: Southern Association of Colleges and Schools Commission on Colleges, 2018), 176.

[19] *Air University Quality Enhancement Plan, 2009–14: "Cross-culturally Competent Airmen"* (Maxwell Air Force Base, AL: Air University, 2009).

Many faculty members thrust into administrative positions focus on what they know—curriculum. However, SACSCOC emphasizes that the crux of a QEP is the ability to demonstrate enhanced student learning. Accreditors, it turned out, were nearly as metrics focused as the military. So, I reviewed every existing assessment instrument I could get my hands on, even meeting with some of their creators. Alas, none fully met the QEP's needs. So, in the end, I punted, proposing four outcomes: knowledge (declarative), skills (behaviors and/or procedural knowledge), attitudes (a controversial inclusion justified by the literature), and application (in novel contexts). This provided a conceptual scaffold for the undertaking and allowed academic units to select the elements that best fit their programs. Just as significantly, it permitted us to adopt a grab bag of assessment techniques.

General Lorenz submitted the 99-page plan to SACSCOC in January 2009.[20] Like all good military plans, it began with a vision for "cross-culturally competent Airmen of all ranks and occupational specialties" and a mission to "create and implement a scientifically sound and institutionally sustainable plan to develop and assess cross-cultural competence across Air University's continuum of education."[21] Every word was carefully chosen to boost the plan's chance of success despite deep-seated resistance to teaching about culture using the QEP's general approach rather than area studies.[22]

In time, I came to fully appreciate the wisdom of Sir Basil Liddell Hart's quip "that the only thing harder than getting a

---

[20] *Air University Quality Enhancement Plan, 2009–14.*
[21] "AU's Cultural Education Efforts on Track, Growing," Maxwell Air Force Base, 16 December 2011.
[22] This preference was not surprising, given that the civilian faculty consisted primarily of historians, international relations scholars, and political scientists, many of whom had benefited from area studies programs in graduate school.

new idea into the military mind is to get an old idea out."[23] The possibility of failure was very real, in my opinion. Nevertheless, the SACSCOC peer reviews described the plan as "pioneering," "broad," "ambitious," and "state of the art." They concluded, "We know of no other U.S. higher education institution, military or otherwise, that has embarked on a plan of this magnitude and we commend Air University for its visionary QEP."[24] No pressure!

The initiative's scale was daunting: the deans of six academic units, from the Air War College to the Community College of the Air Force, had signed on. In addition to developing a curriculum tailored to each and assessing students' learning, the plan called for faculty development and expanding library resources. Implementing the QEP would require a team. Thankfully, in addition to funding, the chief of staff's special initiative included manpower.[25] From these billets, I fashioned an academic department tailored to the 3C model we had adopted. I argued vigorously that it needed to be interdisciplinary, an approach Henk came to embrace. In addition to anthropologists, we recruited a talented group of scholars from the fields of communication, geography, and political science (several of whom contributed to this volume). Some of us were also regionalists, but the department lacked deep expertise in the Middle East and Central Asia. We also recruited a series of industrial/organizational psychologists, whose contributions to measuring student learning across the four outcomes were critical and enriched our approach to teaching 3C.

The work became my calling. Dan Henk's initial vision and unbridled optimism inspired me to become an advocate

---

[23] Basil H. Liddell Hart, *Thoughts on War* (London: Faber & Faber, 1944).
[24] Southern Association of Colleges and Schools, Commission on Colleges (SACSCOC), *Report of the Reaffirmation Committee: Air University Quality Enhancement Plan* (Maxwell Air Force Base, AL: Air University, 2009), 36–37.
[25] *Manpower*, the U.S. military's gendered term for positions, billets, or lines, is used here, rather than *personnel*, DOD's gender-neutral term for the individuals who serve in those roles.

for anthropology's engagement with the military. Publicly, the discussion focused primarily on the Human Terrain System (HTS). I was not alone in questioning that program, both in terms of ethics and efficacy.[26] I came to see the QEP as the converse of HTS: instead of supporting operations, we were building institutional capacity. Rather than providing a handful of external "experts" to deployed Army units, we were developing many generalists (and a handful of specialists) across the entire Air Force.[27] Yet, HTS stubbornly remained the focus of public and academic debate. The American Anthropological Association's unwillingness or inability to address these important distinctions led me to shift my affiliation to the Society for Applied Anthropology. Eventually, I disengaged from the debate altogether, seeking instead to dedicate myself to more productive ventures.

I increasingly had opportunities to engage with Department of Defense-level efforts regarding culture.[28] Unfortunately, due to institutional prerogatives and individual personalities, experts' contributions rarely resulted in coherent or useful policies. I found greater success within the Air Force, where decision makers were more accessible and—for a time—receptive to academic expertise.[29] However, even in these quarters there was an inexorable pull toward "the way

---

[26] See, for example, Robert Albro et al., *Final Report on the Army's Human Terrain System Proof of Concept Program* (Arlington, VA: American Anthropological Association [AAA] Commission on the Engagement of Anthropology with the US Security and Intelligence Communities, 2009); and Ben Connable, "All Our Eggs in a Broken Basket: How the Human Terrain System Is Undermining Sustained Cultural Competence," *Military Review* (March–April 2009): 57–64.

[27] HTS highlighted the military's preference for "buying" solutions rather than "becoming" a different kind of organization. Kerry B. Fosher, "Practice Note: Defense Discourses," *Anthropology News* 49, no. 8 (2008): 54–55.

[28] Daniel P. McDonald et al., *Developing and Managing Cross-Cultural Competence within the Department of Defense: Recommendations for Learning and Assessment* (Washington, DC: Defense Language Office, 2008).

[29] *Air Force Culture, Region, and Language Flight Plan* (Washington, DC: U.S. Air Force, 2009).

we've always done it": high-level foreign language learning, reductionist predeployment training, and area studies. This only redoubled my commitment to infusing military education programs with generalizable cultural knowledge, skills, and attitudes.

By 2014, the QEP had grown from 6 to 12 academic programs. The Culture and Language Center was offering two online courses for enlisted airmen, an executive education course for generals deploying overseas, numerous lessons in professional military education programs, and many electives for graduate degree programs. Initially, the Center's predeployment training materials applied the QEP's approach to shorter culture-specific interventions, providing our efforts even greater reach.[30] Gradually, we even infused career-long foreign language learning with elements of 3C.[31] Behind these efforts was an enormous amount of administration: hiring and managing new faculty, securing and executing budgets, drafting reports and briefing leaders. It was unglamorous but necessary work that enabled the QEP and the center to grow and thrive.

I documented the successes of Air University's QEP for SACSCOC in a 10-page impact report.[32] Unlike the kickoff, the QEP ended with a whimper, not a bang. The report was accepted without any deficiencies—or plaudits. The university was under new leadership—3C was no longer a priority. Henk had retired and a series of less-visionary leaders were hired to oversee the center, tipping the delicate balance be-

---

[30] See, for example, "Air Force Culture and Language Center: Expeditionary Culture Field Guides," AirUniversity.AF.edu, accessed 22 March 2021; and Ryan Haase, "Six Indispensable Apps for Business Travelers," *Wall Street Journal*, 30 August, 2018.

[31] See, for example, "Air Force Culture and Language Center: Language Enabled Airman Program," AirUniversity.AF.edu, accessed 22 March 2021; and "The Air Force Culture and Language Center's Voices of LEAP," YouTube, 6 February 2019, 3:20 min.

[32] *Air University Fifth-Year Interim Review Impact Report of the Quality Enhancement Plan* (Maxwell Air Force Base, AL: Air University, 2015).

tween military and academic decidedly toward the former. I returned to the Office of Academic Affairs and was given a new, challenging portfolio—faculty affairs. Absent the vehicle, vision, and leadership, the culture-general approach made possible by the QEP diminished, but thankfully did not disappear completely.

## In Retrospect
*Successes and Shortcomings*

Although the QEP focused on teaching 3C across Air University, housing the initiative—and the faculty who implemented it—in the Culture and Language Center was critical. Initially, some senior leaders had lobbied to dole out new faculty billets to existing academic programs and schools. Dan Henk's ability to retain administrative control of the team allowed us to tailor the academic expertise, provided a sense of community, fostered innovation, and promoted sharing of best practices. None of this would have been possible had General Lorenz not decided to experiment with university-wide departments "based upon common faculty expertise."[33]

As important was the QEP's colocation with efforts to teach foreign languages. The presence of a Language Training Detachment enriched teaching and, along with the creation of the Language Enabled Airman Program (LEAP), eventually provided opportunities to apply culture-general concepts. In hindsight, I wish we had pursued these opportunities earlier and more vigorously. However, I am gratified such efforts have grown in recent years.[34] The center also provided support services, access to policy makers, and funding. Until measures and requirements exist for culture—whether specific or general—foreign language will continue to receive higher levels

---

[33] James L. Fisher et al., *Air University Review: February 2007–April 2007* (Venice, FL: James L. Fisher, Ltd., 2007), 13.

[34] Jasmine Bourgeois, "AFCLC, Air University's First Virtual LREC Symposium Draws Thousands of People," *Air Force News*, 21 October 2020.

of resourcing. The QEP's ability to benefit from these monies was essential and is a key lesson.

Similarly, while the QEP was an educational initiative, linking it to predeployment training allowed for cross-pollination of ideas and synchronization of efforts. Some faculty and administrators were uncomfortable with this, lest training diminish the academic rigor of graduate programs. Peer reviewers, on the other hand, recognized the benefits early on, noting that "by combining QEP culture-general education with pre-deployment . . . culture-specific training, the Air Force will have a powerful set of tools to enhance the performance of Airmen in carrying out their responsibilities."[35] Overcoming institutional resistance and ensuring conceptual coherence was easier when translating educational approaches into training than vice versa.[36]

The QEP succeeded in meeting peer reviewers' expectations regarding the assessment of student learning; however, success was uneven. Increasing students' knowledge (largely conceptual, informed by the work of Hall, HRAF, and others) and improving their skills (specifically, communicating, negotiating, and relating across cultural differences) was relatively straightforward. The QEP produced impressive results. This should be treated as a preliminary step toward establishing metrics for 3C. Changing students' attitudes and demonstrating the ability to apply learning in novel contexts were more challenging. Measures were elusive, simulations were challenging, and results were often disappointing. Further, given the educational nature of the QEP, we never sought to assess students' performance when deployed. Did their behavior change? Did it make a difference? Clearly, much work remains.

In addition to measuring proficiency, the Department of

---

[35] SACSCOC, *Report of the Reaffirmation Committee*, 37.
[36] Ironically, low-tech printed training materials, such as AFCLC's *Expeditionary Culture Field Guides*, tended to be better received than more expensive ones such as avatars, games, and the like.

Defense's foreign language community has mastered the art of coding positions to require listening, reading, and/or speaking abilities at particular levels. The military assigns specialized personnel to fill these billets, and teaching them is the lifeblood of organizations like DLIFLC. The QEP's contributions on this front were more meager, given the general-purpose student body, transferable approach, assessment challenges, and educational focus. Still, some specialized communities—from foreign area officers to senior commanders—saw the benefits of 3C and established connections. Yet, without a clearer and larger "demand signal" codified in the military's complex human capital system(s), these advances are unlikely to be numerous or permanent. The QEP should be treated as a step in this process, not a stand-alone initiative vulnerable to financial pressures and bureaucratic whims.

Perhaps the most enduring contribution the QEP made was to permit the hiring of so many diverse and gifted faculty members. As Kerry Fosher notes in her chapter, charlatans proliferated when funds started flowing in the defense culture community. However, unlike contractors, who could self-proclaim themselves "culture experts" to unsuspecting contract managers, we recruited and selected academics using the same rigorous peer review processes as civilian universities. Moreover, unlike contractors, faculty occupied billets, providing them greater job stability and, after 2016, the possibility of academic tenure. Military leaders often saw these processes as slow and inefficient, but the results spoke for themselves. I am particularly proud of how many civilian women academics we hired, though the university was less successful at retaining some of them, due largely to quality-of-life issues. Still, many of these scholars, like the editors of this volume, have gone on to serve at other military education institutions, highlighting the U.S. armed forces' growing and interconnected network of academic experts on culture.

The culture-general approach I proposed in the white paper and fleshed out in the QEP proposal stood the test of time.[37] It was far from perfect or complete though. I wish I had pushed sooner for more balance between culture-general learning and application in specific places. Similarly, we could have developed the regional expertise of our faculty more deliberately, to promote application in particular regions, and help assuage the institution's deep-seated preference for area studies. Furthermore, while the model emphasized cognitive, behavioral, and affective aspects of culture, I sincerely regret not having included the material domain.[38] My subsequent teaching about cultural heritage and cultural property protection was successful but too late for inclusion in the QEP, marking a missed opportunity at the institutional level.[39] Finally, faculty worked diligently to ensure 3C was tailored to address the U.S. Air Force's needs—which are different from those of other Services—though most efforts focused at the tactical and operational levels.[40] I believe we were too slow and timid in defining 3C at the strategic level, resulting in occasional frustration and pushback from some of the senior officers we sought to recruit as advocates.

Curriculum is not static though. Many new lessons and courses were developed and revised during the QEP. Unfortunately, since the pace of change is faster in military education than at most civilian institutions, after the plan concluded in

---

[37] Selmeski, *Military Cross-Cultural Competence*; and *Air University Quality Enhancement Plan, 2009–14*.

[38] See, for example, Laurie Rush ed., *Archaeology, Cultural Property, and the Military* (Woodbridge, UK: Boydell Press, 2010).

[39] Brian R. Selmeski, "Bridging the Gap: Synchronizing Material and Behavioral Culture Programs" (presentation, Annual Meeting of the Archaeological Institute of America, San Antonio, TX, January 2011).

[40] For example, the Air Force, unlike other Services, regularly deploys individuals or small teams, rather than medium-size units like brigades or regiments. In addition, while airmen's deployments are often shorter than those of their ground and sea counterparts, they frequently occur with less advance notice. Moreover, a single deployment may take an aircrew to multiple countries. These and other factors meant we could often adapt approaches to cross-cultural learning from other Services, but we could rarely adopt them wholesale.

2014, many lessons and courses were eliminated or reverted back to the traditional area studies approach. Nevertheless, the longer elements of the QEP's curriculum have endured, the more frequently former students have become their best advocates. They have deployed—usually multiple times—and come to their own conclusions about the need for cross-cultural learning. They seek out relevant courses, advisors, and readings at Air University. They take the classroom lessons with them when they go to the field. They reach back to professors and advisors with observations and suggestions. The resistance I encountered in 2007 has softened, suggesting that years of overseas deployments have changed the attitudes of military personnel in general, and airmen in particular, regarding the importance of culture.

Two noteworthy curricular successes were Introduction to Culture and Introduction to Cross-Cultural Communication, lower-division undergraduate courses for enlisted airmen. Whereas some graduate programs participating in the QEP added or modified one or two lessons, these were 40+ contact hour classes. As a result, they served as test beds for curricular materials used in other education (and training) venues, as well as assessment techniques, such as situational judgment tests. Further, because they were delivered online for credit toward the associate in applied science degree, thousands of airmen voluntarily took them each year.[41] This brought balance to what could easily have been an officer-centric educational effort and developed a deeper reservoir of expertise in the much larger enlisted force. I only regret that we were not able to develop the additional courses necessary to establish a concentration in 3C for the Community College of the Air Force's 70-odd degree programs.

As noted earlier, Henk (for the center) and I (for the QEP) spent a disproportionate amount of our time and energy on

---

[41] In academic year 2013–14, 2,200 students enrolled in Introduction to Culture and 930 in Introduction to Cross-Cultural Communication.

administration. This required the ability to understand, navigate, align, and satisfy three systems: government, military, and academic. Eventually, I came to see that in many ways our challenges were as cross-cultural as those our students faced when they deployed. Yet, as necessary as this work was to achieve success at scale, it also had professional consequences for me. Less time in the classroom is a misfortune that unites academic administrators of all stripes. However, close contact with students—accomplished military professionals—is one of the great privileges and benefits of teaching at an institution like Air University. Consequently, while directing the QEP, I had fewer opportunities to test my ideas with military students, learn from them, and develop new insights. On a personal level, it was difficult to watch the talented group of faculty we had assembled teach while I tended to paperwork—no matter how necessary it was. Eventually, I built my portfolio of lectures, exercises, lessons, and courses; however, the process was painfully slow.

Likewise, during these years, I had precious little time to conduct new ethnographic research or even to stay current in my other areas of interest. My "scholarly record" came to be dominated by conference presentations (then, increasingly, serving as discussant), policy documents, and an occasional book chapter, but precious few peer-reviewed publications. When I did write (or speak), the controversies around HTS often compelled me to address or offer lengthy prologues about professional ethics.[42] In the charged atmosphere of anthropology during those years, it increasingly felt like a fool's errand. I became envious of my friends from graduate school as they conducted fieldwork every summer, published regularly, earned tenure, took sabbaticals, published more, and estab-

---

[42] Brian R. Selmeski, *Navigating the Slippery Slope: Balancing the Practical Benefits, Ethical Challenges, and Moral Imperative of Security Anthropology*, Centre for Security, Armed Forces and Society Occasional Paper Series (Kingston, ON: Royal Military College of Canada Centre for Security, 2007).

lished themselves academically. I too worked at a university, though I often felt like I labored in a parallel higher education universe.

Finally, the QEP was selected, funded, and ultimately succeeded in large part because it resonated with various audiences: military and academic. Yet, this came with a price. Within Air University, I was pulled between the Office of Academic Affairs, Culture and Language Center, and academic programs. I worked in a permanent state of liminality; a "know-nothing administrator" in the eyes of some faculty and a naive faculty member in the eyes of "real administrators." This extended beyond the university as well; for years, Henk sent me off to educate, recruit, translate, and influence people and processes across the Department of Defense. Was I a professor or a program manager? An academic or a bureaucrat? Managing these roles and others' impressions of me was critical to the QEP's success, yet personally exhausting.[43] As I have noted elsewhere, serving many masters was a mixed blessing.[44] The possibilities were plentiful and the results gratifying. However, meeting all the competing expectations was ultimately impossible, so some opportunities were missed. Still, it was an honor to be part of this effort; I could not have imagined such an exciting and meaningful opportunity when Henk first invited me to dinner in 2006.

## Conclusion
*Recommendations, Not Recipes*
In the final analysis, I am extremely proud of what the QEP accomplished and contributed to military culture efforts. In less than a decade, we built a team of academic practitioners,

---

[43] Erving Goffman, *The Presentation of Self in Everyday Life* (Garden City, NY: Doubleday, 1959).
[44] Brian R. Selmeski, "Managing a Mixed Blessing: Achieving Educational Success While Serving Many Masters" (presentation, Annual Meeting of the Southern Association of Colleges and Schools Commission on Colleges, Atlanta, GA, 3–6 December 2016).

established a new concept, infused it into military education at scale, and demonstrated how well students had learned. In short, we provided the U.S. Air Force a new and previously unimagined capability.

I hope this brief synopsis provides those involved in current efforts some historical context and those who will launch future efforts an easier starting point than Dan Henk, Kerry Fosher, and I encountered in 2006. It should not be treated as a checklist though. Some of the elements—personalities, agendas, and coincidences—cannot be replicated. Others—from online credit-bearing courses for enlisted personnel and close collaboration between training and education as well as foreign language and cross-cultural efforts—could be adapted more easily. Most importantly, I hope those who work on these issues will continue addressing the shortcomings and lacunae I have identified, particularly measurement, requirements, and strategy, in the years to come. For, despite the rapid decline in the popularity of counterinsurgency, senior military leaders today are increasingly emphasizing the relevance of "culture and language" to the reigning strategic paradigm of "great power competition."[45]

---

[45] Barbara M. Barrett, "Great Power Competition," as quoted in Bourgeois "AFCLC, Air University's First Virtual LREC Symposium Draws Thousands of People," 3.

CHAPTER SEVEN

# Bridging the Social Science Research-to-Practice Gap

*by Allison Abbe, PhD*

## Introduction

My work on sociocultural research in the U.S. Army was the result of two early career detours. Immediately after graduate school, I taught at George Washington University in Washington, DC, for two years and hoped to go on to another academic position, following the model set by my graduate school mentors. At the same time, I was open to alternative options in government, having worked for civilian scientists and engineers in high school as an intern with the Army Corps of Engineers. As a result, after my term ended at George Washington University in 2006, I joined the Army Research Institute for the Behavioral and Social Sciences (ARI). This decision diverted me from a traditional, academic career path, toward a path less well-charted as an applied research psychologist in national security.

I started at ARI with the expectation of contributing to its growing research program on teams, which aligned with my background as a social and personality psychologist and my dissertation research. But in a second career detour, my role shifted after just a few months. ARI was getting questions from the Army Training and Doctrine Command (TRADOC) about how to help soldiers understand the sociocultural dynamics in Iraq. To help ARI better respond to these inquiries, I was soon asked to shift focus and instead lead efforts on culture. Among my new responsibilities was a study that the Center for Army Leadership (CAL) requested on what aspects of cultural understanding and language proficiency were important for officers at different levels.[1] As a relatively junior researcher, I had a surprising level of autonomy to shape the project as I thought best. As I explored the research literature, it was apparent that the approaches to cultural training taking root at the time were inconsistent with the research evidence. Although useful, the "facts about Iraq" typical of predeployment training at the time were insufficient to prepare personnel for the local dynamics they encountered in the operational environment. Research pointed more toward the development of culture-general skills and abilities that would enable personnel to learn about and navigate the culture on their own.

The study for CAL included a workshop with academics and practitioners and produced several technical reports on intercultural skills and training. The report on cross-cultural competence seemed to resonate the most within the defense culture community, resulting in a number of invited conference presentations and participation in enterprise-level working groups on training and education, discussed in the next section.[2] However, the least read of these reports highlighted, in my view, some of the more interesting interdisciplinary research questions that still seem to be unanswered. The potential causal impact of foreign language training on learning

---

[1] CAL was renamed the Center for the Army Profession and Leadership in 2019 when it absorbed West Point's Center for the Army Profession and Ethic.

[2] Allison Abbe, Lisa M. V. Gulick, and Jeffrey L. Herman, *Cross-Cultural Competence in Army Leaders: A Conceptual and Empirical Foundation* (Arlington, VA: U.S. Army Research Institute for the Behavioral and Social Sciences, 2007).

additional languages and intercultural skills was, and continues to be, an issue fraught with assumptions and in need of testing with rigorous research designs.[3]

Although there were no follow-up studies at ARI on the foreign language issues, the study for CAL led to other opportunities in two parallel lines of activity that did not intersect as often as I would have liked. One path led to supporting training and professional development, the other to research and development. Along both paths, I had the good fortune to support both Army strategic initiatives as well as broader defense efforts.

## Training and Professional Development
During the study for CAL, I was invited to participate in the development of the *Army Culture and Foreign Language Strategy* led by TRADOC.[4] Two notable features of this strategy were its intended emphasis on culture over foreign language for general-purpose forces and its level of detail, which probably posed an obstacle in its delayed publication as well as in its implementation. I also participated in related defense-wide efforts, led by what was then the Defense Language Office in the Office of the Secretary of Defense. These collaborations were the most rewarding aspect of the research. Working with other social scientists and program managers to apply existing research in shaping the strategic direction of military professional development was a unique opportunity, one that I would never have experienced as an early career scientist in a traditional academic setting. My hope for these efforts was that they would help institutionalize cultural training and education throughout a servicemember's career. The culture-general skills that literature showed to be most predictive of success

---

[3] Allison Abbe, *Transfer and Generalizability of Foreign Language Learning* (Arlington, VA: U.S. Army Research Institute for the Behavioral and Social Sciences, 2009).

[4] *Army Culture and Foreign Language Strategy* (Washington, DC: Department of the Army, 2009).

working abroad were not trainable in a predeployment training cycle. These skills and abilities took time to develop and were probably not achievable within a single training event or rotation to a combat training center. Thus, ensuring that soldiers received exposure and development at different phases of their careers was critical, from officer precommissioning to functional training to predeployment.

Working with the Army organizations responsible for overseeing and delivering cultural training and education in turn helped guide further research projects at ARI, highlighting key knowledge gaps. Projects included comparing the validity, reliability, and utility of different measures of intercultural competence and identifying the cultural competencies needed by tactical-level leaders.[5] Contract funding enabled other projects on intercultural perspective taking and developing assessment tools for cross-cultural competence specific to military interactions.[6] At the time, it was not clear how long the expanded U.S. commitment in Iraq would continue. Thus, to avoid the same institutional amnesia after Vietnam, it seemed important to maintain a focus on intercultural skills and abilities relevant in any country and supportive of other missions, not only in counterinsurgency or counterterrorism.

The collective efforts in getting evidence-based culture-

---

[5] Allison Abbe, David S. Geller, and Stacy L. Everett, *Measuring Cross-Cultural Competence in Soldiers and Cadets: A Comparison of Existing Instruments* (Arlington, VA: U.S. Army Research Institute for the Behavioral and Social Sciences, 2010); and Allison Abbe and Jessica A. Gallus, *The Socio-Cultural Context of Operations: Culture and Foreign-Language Learning for Company-Grade Officers* (Arlington, VA: U.S. Army Research Institute for the Behavioral and Social Sciences, 2012).

[6] Michael J. McCloskey et al., *Assessing the Development of Cross-Cultural Competence in Soldiers* (Arlington, VA: U.S. Army Research Institute for the Behavioral and Social Sciences, 2010); Joan R. Rentsch et al., *Conceptualizing Multicultural Perspective Taking Skills* (Arlington, VA: U.S. Army Research Institute for the Behavioral and Social Sciences, 2007); and Joan R. Rentsch, Ioana Mot, and Allison Abbe, *Identifying the Core Content and Structure of a Schema for Cultural Understanding* (Arlington, VA: U.S. Army Research Institute for the Behavioral and Social Sciences, 2009).

general approaches incorporated into strategic guidance, training, and education had some success, at least partly due to collaboration across disciplines and organizations. It helped that multiple voices were communicating reinforcing messages about the importance of culture-general skills in general-purpose forces, though in different forms and disciplinary language. It was critical that these voices were positioned within their organizations to make an impact. Timing was also a factor, in that the Army's *Counterinsurgency*, Field Manual 3-24, had been published just prior to the strategy development efforts and incorporated cultural considerations.[7] This alignment with doctrine was an important factor.

One key limitation was the scale of the enterprise. The geographic and organizational dispersion of Army schools made it difficult to align approaches. For several years, from 2007 to 2011, the TRADOC Culture Center hosted an annual culture summit in the spring. The summits were very helpful for sharing information, supporting collaboration among geographically dispersed centers, and supporting research-to-practice transitions.[8] Unfortunately, those summits were a casualty of government-wide restrictions on conferences. An investigation into excessive conference spending by non-DOD government agencies coincided with an executive order on promoting more efficiency in government spending.[9] As a result, strict guidance from the Office of Management and Budget limited government conferences and personnel travel, inhibiting knowledge sharing across the DOD.

---

[7] *Counterinsurgency*, Field Manual 3-24 (Washington, DC: Department of the Army, 2006).
[8] Amy Sunseri, "Culture Summit Brings Nations Together, Promotes Understanding," U.S. Army, 9 March 2011.
[9] Eric Katz, "Looking Back at the GSA Scandal: Did the Administration Overreact?," *Government Executive*, 26 January 2015; and Exec. Order 13589, 76 F.R. 70861 (9 November 2011).

## Science and Technology (S&T) Programs

At the same time, I collaborated with organizations outside of ARI that were funding external academic and private sector research, aiming to build more support for basic research on culture. The behavioral and social science evidence base to inform training and education was limited by a lack of previous investment. The Department of Defense's preference for materiel solutions to problems had manifested in decades of neglect of defense research and development funding for the behavioral and social sciences. Since World War II, technological innovation and superiority have been central to U.S. military strategy and dominate defense research and development investments as a result.[10] Within the Army, the Army Research Office (ARO) had the lead on basic research but, at the time, had no behavioral or social scientists on staff. As a result, ARO looked to ARI for input as the pressing sociocultural research needs became apparent.

ARI colleagues and I worked with ARO to develop a multidisciplinary university research initiative (MURI) on modeling intercultural collaboration and negotiation in 2008, which aimed to advance basic research. Previous research in organizational settings had been primarily cross-cultural, comparing norms, values, and interactions *within* different cultures. Cultural dimensions frameworks are an example of such cross-cultural research, such as the Geert Hofstede or GLOBE dimensions.[11] Research on intercultural interactions in defense contexts was far more limited. Whereas cross-cultural research explores how cultures differ from each other, intercultural research focuses on the dynamics of interaction

---

[10] Adrian R. Lewis, *The American Culture of War: The History of U.S. Military Force from World War II to Operation Iraqi Freedom* (New York: Routledge, 2007); and John F. Sargent Jr., *Defense Science and Technology Funding* (Washington, DC: Congressional Research Service, 2018).

[11] Geert Hofstede and Michael H. Bond, "Hofstede's Cultural Dimensions: An Independent Validation Using Rokeach Values Survey," *Journal of Cross-Cultural Psychology* 15, no. 4 (November 1984): 417–33.

between cultures.[12] ARO's MURI funding enabled awards to two different research teams, one led by psychologist Michele Gelfand at the University of Maryland and one led by computational scientist Katia Sycara at Carnegie Mellon University. These multiyear contracts resulted in dozens of peer-reviewed publications, hundreds of presentations, and multiple research awards, supporting more than 30 graduate students and significantly advancing interdisciplinary intercultural research.[13]

I also had a peripheral role in the Human, Social, Cultural, and Behavioral (HSCB) Modeling Program in the Directorate of Defense Research and Engineering, providing subject matter expertise for a line of applied research on training. However, even with "human," "social," and "cultural" in the program title, the appetite for engineering and computational approaches far exceeded any emphasis on behavioral and social sciences. The initial focus of the HSCB program tended to be on the cultural "other" with limited consideration of the operational context and servicemembers for which the research would be applied. My focus was therefore both to highlight the importance of advancing the behavioral science and to maintain attention on soldiers and servicemembers as the end users and beneficiaries of the research.

More broadly, though, the applied computational approaches would not succeed unless accompanied by investment in the underlying basic behavioral research. Computational models could be invalid and misleading if reliant on behavioral science untested in other cultures or for the intended purposes. In addition, some of the modeling efforts funded by defense programs struck me as too dependent on experts, with insufficient attention to how the modeling tools might

---

[12] John W. Berry et al., *Cross-Cultural Psychology: Research and Applications*, 2d ed. (Cambridge, UK: Cambridge University Press, 2002).

[13] Michele Gelfand, *Final Report: Dynamic Models of the Effect of Culture on Collaboration and Negotiation* (Research Triangle Park, NC: Army Research Office, 2014); and Katia Sycara et al., *Modeling Cultural Factors in Collaboration and Negotiation* (Research Triangle Park, NC: Army Research Office, 2014).

integrate into operations. Would the models be practical for end users who might not be familiar with the underlying social science or with the computational modeling approaches? Also, experts and specialists might be available to support integration of such models at the operational and strategic levels, but at the tactical level, there was no substitute for servicemembers being able to make sense of a dynamic, unfamiliar operating environment for themselves. Insufficient consideration of the soldier or Marine as the end user was consistent with a broader pattern of neglect of small unit needs.[14] Six years into the HSCB program, the program manager acknowledged that developing "S&T [science and technology] solutions for individual warfighters remains a long-term goal."[15]

## Bridging the Divide

At a personal level, my own limited success in supporting evidence-based and research-informed approaches was in a bridging role between researchers and practitioners. In my experience, practitioners and managers are often interested in and open to social science but do not have time to wade through academic journals. The jargon and methods need translation into actionable principles. Unfortunately, the typical incentive structure for academic researchers can have the unintended effect of researchers metaphorically throwing their research over the fence to practitioners. Incentives in academic institutions tend to encourage researchers to write primarily for other researchers and to secure resources to do more research, leaving a gap between the science and its transition into practice. When researchers experience little professional benefit from having a practical impact on organizations, practice suffers from the disconnect. Practical impact

---

[14] *Army Transformation: Hearing before the Committee on Armed Services*, 108th Cong., 2d Sess. (Robert Scales, testimony, 21 July 2004).
[15] Capt Dylan Schmorrow, "Sociocultural Behavior Analysis and Modeling: Technologies for a Phase 0 World" (brief, Office of the Assistant Secretary of Defense [Research and Engineering], 6 March 2013).

requires building relationships, working with organizations to provide input to policy and to review curricula, being externally facing and participating in meetings—activities that do not increase one's peer-reviewed publications, bibliometrics, or grant funding.

For me, performing that bridging role as an Army civilian had some unique challenges that I had to learn to overcome through observation and trial and error. For example, I found I needed to broaden my disciplinary horizons. As an applied psychologist, my role was to help solve practical organizational problems, drawing on my technical expertise and disciplinary perspective. But on the topic of culture, a lot of disciplines have something to contribute, and no discipline has a monopoly on useful knowledge. Psychology uses a particular set of theories and methods that have both strengths and limitations and, for me, gaining greater familiarity with anthropology, sociology, communications, and historical perspectives was helpful in understanding the limits of my own disciplinary perspective and how to complement other efforts.

I also had to learn to expand my communication skills. Although my graduate studies gave me some exposure to communicating outside my discipline, my program and many others do not adequately prepare students to talk to policy makers, practitioners, or the public. It requires simplifying one's writing and presentation style, not because the audience lacks sophistication or intelligence, but because they have enormous demands on their time and attention. Being able to communicate complex ideas and research findings clearly and concisely is a valuable skill, which social scientists need to practice and refine along with their research skills. It is especially important in social science because, although the language of the discipline may be more accessible than that of other sciences, lay audiences often lack an understanding of the methods and foundations for scientific claims.

One key shift needed in communicating with practitioners, versus with other scientists, is to focus on what is known as opposed to unknown. Many presentations by academic scientists ended with some variant of the statement, "More research is needed." This response invariably frustrated the practitioners in the room. They wanted actionable conclusions, not a proposal for additional research. The scientists often wanted to think about further advancing the science, recognizing that conclusions may be tentative or incomplete and multiple studies are necessary. In contrast, for practitioners, emphasis on the scientific consensus to date was more helpful, though often less interesting to the scientists, whose interests and career advancement depend on the pursuit of unanswered questions.

Another issue that I encountered repeatedly is that organizational time lines and the deliberate pace of the scientific process are often not compatible. Budget cycles and senior leader decisions do not wait for the results of a well-designed scientific study with statistically powerful sample sizes. To have real organizational impact when the organization needs it, you may need to provide input when the level of scientific uncertainty is uncomfortably high. Offering the state of knowledge at the time is still important, while acknowledging the limitations of the science and the potential for conclusions to change when further study is complete. Unfortunately, there are many examples of social science research being applied prematurely or inappropriately (e.g., coercive interrogation techniques based on learned helplessness).[16] Having social and behavioral scientists engaged at multiple levels of decision making may help prevent such problems, and professional organizations are shifting toward more such engage-

---

[16] Laurence Alison and Emily Alison, "Revenge versus Rapport: Interrogation, Terrorism, and Torture," *American Psychologist* 72, no. 3 (2017): 266–77, https://doi.org/10.1037/amp0000064; and Martin Seligman, "The Hoffman Report, the Central Intelligence Agency, and the Defense of the Nation: A Personal View," *Health Psychology Open* 5, no. 2 (July–December 2018): 1–9, https://doi.org/10.1177/2055102918796192.

ment, though engaging on defense issues continues to be controversial in some subdisciplines.[17]

Social scientists will bridge the research-practice divide only to the extent that their employers and professional disciplines reward it. In my case, bridging activities certainly cut into my research productivity, but I was lucky to have supportive supervisors at ARI who allowed me a great deal of autonomy. I had wide latitude in engaging externally, with my supervisors supporting many travel requests to participate in external meetings and events without imposing burdensome internal reporting requirements. Moreover, they supported my promotion while at ARI, a very tangible metric of what the organization values.

## Organizational Challenges and Obstacles

Despite some early success, the impact of these efforts in the Army was not as pervasive or as lasting as I had hoped. Some of the biggest obstacles were applicable to any organizational change effort in the Army. For example, senior leader turnover in the Army is a major barrier to successful institutional change. The short duration of senior leader assignments, often just a year or two, means that organizational change efforts often start off strong, as new leaders want to make their mark on the organization, but do not survive the transition to a successive leader.[18] In the case of the Army's cultural training and development, senior leader advocacy was limited at the outset. As a result, culture efforts in the Army never had sufficient momentum. Leader turnover also meant that the bottom-up efforts could not take root, as change initiated at

---

[17] Craig R. Fox and Sim B. Sitkin, "Bridging the Divide between Behavioral Science and Policy," *Behavioral Science and Policy* 1, no. 1 (2015): 1–12; and Stephen Soldz, Bradley Olson, and Jean Maria Arrigo, "Interrogating the Ethics of Operational Psychology," *Journal of Community & Applied Social Psychology* 27, no. 4 (2017): 273–86, https://doi.org/10.1002/casp.2321.

[18] Margaret C. Harrell et al., *Aligning the Stars: Improvements to General and Flag Officer Management* (Santa Monica, CA: Rand, 2004).

lower levels had to start over so frequently, persuading a new leader of the value and relevance of culture.

An additional obstacle to incorporating evidence-based social science approaches was the established organizational structure. Budget and strategy were heavily influenced by organizations that already had relatively large resources and more influence. In contrast, social and behavioral scientists in the Army were few and distributed, and they were not well-positioned to influence organizational direction and budget decisions. Instead, well-established communities, such as the foreign language and engineering communities, tended to have more influence and, whether intentionally or implicitly, they often acted to defend organizational interests and promote solutions that fit neatly within their scope of responsibility.

Organizational challenges also inhibited the bridging of research-to-practice transitions. This problem is not unique to the social and behavioral sciences; getting basic and applied science through the "valley of death" to applications that can be tested and used by practitioners is a well-recognized problem in the development of defense technologies.[19] Organizations that fund basic and applied research (6.1 and 6.2 funding in DOD terminology) may not have the personnel, budget authority, or incentives needed to facilitate transition to practice.[20] In the social and behavioral sciences, the transitions are often much simpler than for technologies, as the goal may be as modest as adapting a theory or set of concepts for policy or practitioner use. The time needed to transition social science should therefore be shorter than for technology innovations, but it still requires deliberate effort. Defense

---

[19] National Research Council, *Accelerating Technology Transition: Bridging the Valley of Death for Materials and Processes in Defense Systems* (Washington, DC: National Academies Press, 2004), https://doi.org/10.17226/11108.

[20] Donna Fossum et al., *Discovery and Innovation: Federal Research and Development Activities in the Fifty States, District of Columbia, and Puerto Rico* (Santa Monica, CA: Rand, 2000), appendix B, https://doi.org/10.7249/MR1194.

research organizations need to allow personnel the time and space to explore transition paths, activities that require different impact assessment than the usual scientific productivity metrics. For these transitions, research conferences and other knowledge-sharing mechanisms are critical, and must also be accompanied by further research-practitioner collaboration to ensure the application of social science is appropriate, useful, and practical. If such exchanges are minimal or absent, then social science investments such as those of the Minerva Research Initiative may represent missed opportunities.[21]

Other challenges involved the intersection between sociocultural issues and the Army's own organizational culture. The Army's continuing ambivalence about culture reflected long-standing conflicts within its organizational identity—the kind of wars the Army prefers to fight versus the kind of roles and missions it is frequently asked to take on. Although commanders at the tactical and operational level in Iraq and Afghanistan adapted to the demands of the fight, leaders at the strategic level were slower to shift.

The Army's recurring neglect of sociocultural aspects of conflict is puzzling for several reasons. With good neighbors on our borders to the north and south and oceans to our east and west, the United States is unlikely to face the need to fight on our own soil. Wherever the Army fights, it will happen elsewhere, and the populations in those locations add a complex dimension to any operation. While the Air Force and Navy may be able to fight from a distance far removed from both the adversary and civilians in the area, ground forces cannot expect to fight at a distance. Second, the American way of war is to fight with allies. In addition to enhancing interoperability, understanding allies and partners' interests,

---

[21] The Minerva Research Initiative, funded by the Office of the Secretary of Defense, has supported social science relevant to strategic national security issues since 2008. Allen L. Schirm, Krisztina Marton, and Jeanne C. Rivard, eds., *Evaluation of the Minerva Research Initiative* (Washington, DC: National Academies Press, 2020), https://doi.org/10.17226/25482.

identities, values, and norms can be a huge asset in integrating them into U.S. efforts. Further, the United States maintains extensive security cooperation relationships and is quite dependent on these relationships for global posture and power projection.

Thus, the Army has a persistent need to incorporate sociocultural perspectives but has proved reluctant to do so in any sustained manner. Countless initiatives have started, but for various reasons have ultimately failed to gain traction beyond a few years. The 09L heritage linguist program, Human Terrain Teams, TRADOC's Culture and Foreign Language Advisors at the Centers of Excellence, the Human Dimension Concept, and regionally aligned forces are examples of organizational adaptations addressing sociocultural issues that have come and gone.[22] The Army continues to reshape its advisory mission, from military transition teams to advise-and-assist brigades to the current Security Force Assistance Brigades, and the high demand for those capabilities will ensure that they persist in some form.[23] However, it is clear from current

---

[22] The 09L designation is an enlisted Army military occupational specialty (MOS) for interpreters/translators who speak both English and a heritage foreign language. Although still in existence, the languages are limited to Arabic, Dari, Farsi, and Pashto. Christopher J. Sims, *The Human Terrain System: Operationally Relevant Social Science Research in Iraq and Afghanistan* (Carlisle Barracks, PA: Army War College Press, 2015); and Amy Alrich et al., *The Infusion of Language, Regional, and Cultural Content into Military Education: Status Report* (Alexandria, VA: Institute for Defense Analyses, 2012). In 2019, the Army replaced the 2015 Human Dimension Strategy with the more inward-looking Army People Strategy, in which references to culture all focus on the Army's organizational culture. An initiative during Gen Raymond T. Odierno's tenure as chief of staff of the Army, cultural preparation for regionally aligned forces has since faded from priority, but the Army's new force generation process adopted in 2020, the Regionally Aligned Readiness and Modernization Model, has potential to revive it. The State Partnership Program within the National Guard may be an exception, now in place for more than 25 years.

[23] Andrew Feickert, *Army Security Force Assistance Brigades* (Washington, DC: Congressional Research Service, 2020).

Army priorities and structure that navigating other cultures continues to be a specialty role or mission, not a core institutional competency.

I would also note that optimism about technology continues to pose an obstacle in sustaining institutional attention to the critical importance of the sociocultural dimension of applying land power. This optimism is defense wide and is not limited to the Army, but the Army suffers from the illusory effects of technology more acutely than the Air Force or Navy, the Services for whom technology and platforms play a more central role. Recurring cycles of technology fervor occur alongside the waxing and waning attention to sociocultural issues. The revolution in military affairs, net-centric warfare, the third offset, and current concepts of highly networked Joint all-domain operations—these concepts reflect assumptions that technology alone may achieve military ends in spite of plentiful experience with thinking, adapting adversaries who refuse to confront the United States in its preferred way of war.[24]

Because the Army continued to see counterinsurgency (COIN) and stability operations as anomalous requirements, some leaders were happy to "get back to" the Army's desired core competencies of large-scale ground combat operations and leave COIN and cultural considerations behind, as also occurred after the war in Vietnam.[25] The perception that culture was a COIN issue may have temporarily made the Army more receptive to social science approaches initially, but it became a disadvantage later. Tying sociocultural considerations so closely to COIN meant that once the COIN focus subsided, so did support for culture-related efforts. Culture is certainly a

---

[24] Christian Brose, "The New Revolution in Military Affairs: War's Sci-Fi Future," *Foreign Affairs* 98, no. 4 (May/June 2019): 122–34.

[25] David Fitzgerald, *Learning to Forget: US Army Counterinsurgency Doctrine and Practice from Vietnam to Iraq* (Stanford, CA: Stanford University Press, 2013).

critical consideration in COIN, but it is no less relevant in an era of great power competition.[26]

With the current focus on Russia and China as our competitors, culture has fallen out of fashion, as Army leaders do not foresee the need for cultural understanding in large-scale ground combat operations. However, if history is any guide, direct conflict with near peer competitors is more likely to happen in a third country or via proxies. In those circumstances, political support is not as simple as *for* or *against*. U.S. forces will need to understand the local population and dynamic attitudes and loyalties. Our actions toward the local population can potentially win them over, maintain neutrality, or drive them toward the opponent. And in contrast to our near peer competitors, allies and partners are critical to U.S. defense strategy, representing a unique American advantage in great power competition.[27]

Ideally, understanding the sociocultural dynamics of allies and adversaries would come well in advance of a major conflict or crisis. Many such efforts are undoubtedly occurring across the Army, though they lack coordination and are not systematically integrated into training and education. Until the organizational culture shifts to include the sociocultural dimension as a fundamental consideration for land power, the Army will continue the cycle of institutional amnesia and rediscovery.

Future efforts will have at least three key sources of help to avoid starting over from scratch: specialist communities of practice, the policy and paper trail, and people. First, special-

---

[26] Jeannie L. Johnson, "Fit for Future Conflict?: American Strategic Culture in the Context of Great Power Competition," *Journal of Advanced Military Studies* 11, no. 1 (Spring 2020): 185–208, https://doi.org/10.21140/mcuj.2020110109.

[27] Matthew Kroenig, "Why the U.S. Will Outcompete China," *Atlantic*, 3 April 2020; and David A. Wemer, "U.S. Joint Chiefs Chairman Makes the Case for Keeping U.S. Troops in Europe," *New Atlanticist* (blog), Atlantic Council, 21 March 2019.

ist communities, such as linguist/interpreters, foreign area officers, civil affairs, and other special operations forces, continue to play an important role and will be a resource for expanding organizational efforts again in the future. Second, the research studies and organizational efforts from the recent COIN era have been more systematically, though not thoroughly, documented than in previous eras. Strategic guidance, policy documents, published research, and conference proceedings provide a richer foundation for future rediscovery. Third, whether intentionally or incidentally, the military Services are retaining some sociocultural expertise on the faculty of professional military education institutions and within research organizations, even if active research has been diverted to other priorities. As of this writing, the institutional forgetting is only in process, and perhaps these three resources will illuminate the path for the social scientists who lead the next rediscovery.

CHAPTER EIGHT

# A Few Things I Know about Culture Programs or Why Nothing Works[1]

*by Kerry B. Fosher, PhD*

## Introduction

To give some context to what follows, I will begin my chapter with an account of my own career path and some reflections on how I think that path shaped my perception of the landscape over time. In 2004, I was at a conference where colleagues encouraged me to attend a session where panelists would be discussing how to integrate culture into U.S. professional military education (PME). At the time, I was working at Dartmouth (now Geisel) School of Medicine in New Hampshire in a position primarily focused on biosecurity and homeland security and had an interest in the military, so I attended. The session was interesting, but I noted the way regional knowledge and culture seemed to be conflated and was struck by the fact that there was no mention of teaching basic social science concepts or skills that military personnel could use in a range of places. Although I was unaware at the

---

[1] With apologies to Marshall Sahlins and Marvin Harris, "Two or Three Things that I Know about Culture," *Journal of the Royal Anthropological Institute* 5, no. 3 (September 1999): 399–422; and Marvin Harris, *Why Nothing Works: The Anthropology of Daily Life* (New York: Simon and Schuster, 1981).

time, that experience alerted me to issues that would be of importance in the coming years.

In the spring and summer of 2006, two things occurred that shaped the rest of my career. The first was my appointment to the American Anthropological Association's commission tasked with examining the ethics of working with the military and intelligence community.[2] My four years on that commission focused my attention on the importance of long-term work in building lines of communication between the military and academia, exposed me to how other disciplines were approaching similar issues, and pushed me to clearly articulate both my ethical decision making and the kinds of work in which I was involved. The second shaping event was an invitation from an anthropological colleague, Dan Henk, at Air University to come to a small meeting he was convening to consider how to build a culture education and training program for the U.S. Air Force. I was intrigued by the challenge and for the six months following the meeting, Brian Selmeski and I contributed time in our off hours to help shape the Air Force's efforts around the idea of cross-cultural competence, including providing education in generalizable, culture-general concepts and skills. I was asked to come to Air University, on temporary loan from Dartmouth, to direct the Cross-Cultural Competence Project, which later grew into the Air Force Culture and Language Center.[3] I agreed, fully intending to return to Dartmouth.

The work at Air University was expansive, thanks to the foresight of Dan Henk who recognized that it was critical to

---

[2] James Peacock et al., *Final Report, November 4, 2007* (Arlington, VA: American Anthropological Association [AAA] Commission on the Engagement of Anthropology with the US Security and Intelligence Communities, 2007); and Robert Albro et al., *Final Report on the Army's Human Terrain System Proof of Concept Program* (Arlington, VA: AAA Commission on the Engagement of Anthropology with the US Security and Intelligence Communities, 2009).

[3] "U.S. Air Force Fact Sheet: Cross-Cultural Competence," Air University, November 2017.

build the Department of Defense (DOD) and international networks of people working on these issues. I traveled a great deal, often home only long enough to do some laundry and get a night's sleep before heading out again. These trips allowed me to connect with a network of military and scholarly colleagues who were working toward similar goals across the U.S. Services and internationally. While our jobs varied, we had overlapping aspirations to improve the culture-related education and training available to military personnel. During the following years, this network was essential in many ways, including conceptualizing programs, shaping policy, determining how to assess effectiveness, and resisting the many pressures to have learning conform to outdated theories and approaches. For example, anthropological theories from the 1940s and 1950s, which emphasized categories of cultural information, such as politics, economy, etc., were a very good fit with how the DOD wanted to build its programs. The fact that these theories had been shown to be invalid and replaced long ago did little to convince DOD organizations that they should not be used. We worked on these issues not only within the confines of our specific positions but also informally, each assisting the others to make progress on broad goals, such as shifting programs to a more solid conceptual basis.

After about six months at Air University, I decided not to return to Dartmouth. This was an incredibly difficult decision personally and one fraught with professional challenges, but I had become invested in trying to create change within the military bureaucracies and was reluctant to drop the work.[4]

---

[4] Kerry Fosher, "Yes, Both, Absolutely: A Personal and Professional Commentary on Anthropological Engagement with Military and Intelligence Organizations," in *Anthropology and Global Counterinsurgency*, ed. John D. Kelly et al. (Chicago, IL: University of Chicago Press, 2010), 261–71; and Kerry Fosher, "Pebbles in the Headwaters: Working within Military Intelligence," in *Practicing Military Anthropology: Beyond Expectations and Traditional Boundaries*, ed. Robert A. Rubinstein, Kerry B. Fosher, and Clementine Fujimura (Sterling, VA: Kumarian Press, 2012), 83–100.

So, I accepted a position as the first command social scientist at Marine Corps Intelligence Activity, where I stayed for three years. I then became the director of research for the Marine Corps' culture center, the Center for Advanced Operational Culture Learning (CAOCL). However, the early experiences working on goals from informal or temporary standpoints shaped how I approached my work throughout my time with culture programs. Rather than thinking about pursuing specific jobs, I was focused on a set of broad, at times perhaps a little hazy, goals that I could pursue from different angles. The jobs were platforms from which to do the work rather than ends in themselves. Of course, this orientation was made easier by the fact that I was not looking for a job when I started and by the astonishingly wide range of DOD jobs available to anthropologists at the time. Even as positions became scarcer, my concern about the potential for ethical challenges in work with the DOD caused me to live in ways that preserved my ability to leave any job.[5] That orientation was a luxury that very few people had, and it was not particularly comfortable for me, but it did allow me to stay focused on goals rather than jobs.

One other aspect of the DOD's culture efforts shaped my approach over the years—its cyclic nature. The last 15–20 years are not the first time the DOD has become enamored with culture and social scientists. In the past, that interest has been ephemeral with attention and funding waning as major combat operations wound down and other programs and

---

[5] Fosher, "Yes, Both, Absolutely," 261–71.

buzzwords gained prominence.[6] Those of us familiar with this history knew it was highly likely that our work was part of another cycle that would end with the closure or weakening of the programs we helped to build.

Like everyone else, I spent time caught up in the work of my specific jobs and the opportunities and obstacles they presented. Yet, my early experiences and my knowledge of the likely fate of culture programs did cause me to focus on broad issues and shared challenges. It also ensured that I kept one eye open for lessons learned that we could pass along to those involved in the next cycle of interest. Some of those lessons are outlined in the remainder of this chapter.

## Lessons Learned

While I do not want this part of the chapter to come across as overly negative, I do want to capture some of the thorny challenges we faced and mistakes we made or narrowly avoided in standing up and maintaining culture programs. I hope the following sections—the first on bureaucratic issues and the second on experts—will be informative to current readers, but also useful to future readers, some of whom may be involved the next time the DOD realizes it needs culture programs.

---

[6] Accounts of past efforts to integrate culture and social scientists into DOD and intelligence programs are available for World War II in David H. Price, *Anthropological Intelligence: The Deployment and Neglect of American Anthropology in the Second World War* (Durham, NC: Duke University Press, 2008); for the Vietnam conflict in Allison Abbe and Melissa Gouge, "Cultural Training for Military Personnel: Re-visiting the Vietnam Era," *Military Review* 92, no. 4 (July/August 2012): 9–17; Seymour J. Deitchman, *The Best-Laid Schemes: A Tale of Social Research and Bureaucracy*, 2d ed. (Quantico, VA: Marine Corps University Press, 2014; original printing by MIT Press, 1976); and for the broader Cold War era in David H. Price, "Cold War Anthropology: Collaborators and Victims of the National Security State," *Identities: Global Studies in Culture and Power* 4, nos. 3–4 (1998): 389–430, https://doi.org/10.1080/1070289X.1998.9962596. Some of these works focus on one discipline—anthropology—but provide insights into the wider spectrum of culture-related efforts.

## Bureaucratic Gravity
*Strategic Communication*
Perhaps the most important lesson learned from this cycle is the need for well-organized strategic communication starting as soon as possible. In the early days of current culture programs, culture was on most military leaders' minds already and there seemed little need to raise awareness. It later became apparent that this was a misperception on our part. Those who were or soon would be in positions to make decisions about culture centers had varying degrees of understanding of what culture centers actually did or the relative value of different types of content and programs. These misunderstandings manifested themselves in later years with the same leaders (or others influenced by them) making decisions based on false understandings of what the programs did. For example, we too often heard the idea that culture was only useful in missions like stability operations. We could and should have done more in the early days to shape people's understanding of the enduring value of culture-related programs.

*The DOD Has a Comfort Zone with Its Own Gravitational Pull*
The DOD's preference for certain types of problem framing and solutions was a consistent challenge. Even when there was initial agreement to establish more scientifically sound and operationally relevant solutions, without constant attention and maintenance, things would gradually slide back toward the comfort zone. The slide was an accumulative process, happening through small changes in editing, budgets, policies, and doctrine that were not appropriately staffed and other seemingly minor actions. A few examples of persistent, low-grade resistance that we experienced included:
- The push for highly structured approaches in education, training, planning, and intelligence, such as simplistic frameworks, checklists, and databases that ended up depicting culture as

something static and made out of predictably interacting parts. This happened despite the fact that contemporary science, as well as the realities in operations, strongly indicated the need for approaches to culture that addressed change and the role of individual human decision making.
- The tendency to view nation-state and geographic combatant command boundaries as the most salient geographic features and insistence on having education, training, and other materials organized along these lines. This was despite broad acknowledgment of the importance of nonstate actors and transregional threats.
- The endless search for a technology-based solution. At various points during my time with the DOD, people have pursued a solution to "the culture problem" through things like databases, network analysis tools, models and simulations, social media mining platforms, mapping software, and, as if in desperation, handheld personal translators. DOD has a vast apparatus for developing and purchasing technology, relatively little for purchasing science, and has difficulty investing in long-term solutions like education, so it was especially difficult to convince people that technology was unlikely to provide a magic bullet without some changes in the thinking of the people using it.

The comfort zone issue is not a problem with a solution, but rather one like the weather. If you know it is going to happen, it is easier to recognize when and where you will need to make time to cope with it.

## The Tyranny of Existing Programs, Policies, and Time

At the time we were starting culture programs, DOD and the intelligence community had fairly robust programs for regional expertise and language. Region was used as a gloss for culture-specific knowledge and there was no concept for transferable culture-related concepts and skills—what we now call culture-general. We did an insufficient job of institutionalizing culture, especially culture-general. Consequently, there was a constant, slow tug to reframe efforts and programs as language and region, which we knew would eventually squash culture out due to the perishability of culture-specific information and the relative scarceness of people with the expertise needed to build and maintain culture-general curricula.[7]

In the first few years of the current interest cycle, there was a fairly robust cross-Service effort among those of us involved with culture programs to influence policy at the DOD and Service levels. A number of us scattered throughout the nascent programs knew culture needed to be addressed more explicitly in high-level policy if it was going to survive over time. It was not easy to carve out time for this policy work, as we were all engaged in building programs, researching approaches to content and format, and delivering training and education to military personnel who needed it immediately.

---

[7] Language programs and policy were a special challenge both because the DOD could produce quantified measures of language capability and because language advocates argued that you learn culture by learning language. We argued that it was necessary but insufficient on three counts. First, learning a language does not make you an expert in the culture-specific details of all the places the language is spoken. Second, language has proven to be one of the more difficult and expensive capabilities to develop and is highly perishable. Third, on the culture-general front, it is quite possible to be fluent in a language and still not have the skills to interact effectively, which was a common enough occurrence that there was a joke about it among senior foreign area officers: "That guy knows three languages and is a jerk in all of them."

We saw some initial successes in shifting policies and structures, such as circling the wagons to get DOD to create a culture office within the larger office that handled regional and language programs. A group of us continued to work with that culture office to develop policy and baseline standards for cross-cultural competence-related learning. Unfortunately, after a while, our ability to pay close attention waned as other work reclaimed our time. That waning attention led to problems. The DOD-level culture office, which still exists as of this writing, struggled. It was unable to get policy issued for many years and ended up involving itself more heavily in training and research than most of the Service culture centers found useful. As of this writing, when culture programs across the DOD are shutting down or being reduced, the office has had only limited success in institutionalizing culture in appropriate policy.[8]

The takeaway is that we all assumed we would somehow get back to paying attention to that higher level of policy. However, apart from reacting to various policy drafts and other actions taken by the DOD-level culture office, we never did. That was a mistake. Policy at higher levels could have driven how requirements were written, which in turn could have provided an incentive for the Services to keep culture centers open and well resourced. Instead, we were left with policies that will likely be allowed to lapse or simply reintegrated into region and language policies with culture gradually fading away.

## *Christmas Trees and Niche Missions*

Right now in the DOD, when a concept gets attention and funding, everyone else tries to associate their programs and initiatives with it—however tenuous the connection may be—like trying to hang ornaments on the new Christmas tree.

---

[8] "Culture: Culture Section," Defense Language and National Security Education Office, accessed 22 February 2021.

Initially, culture was a Christmas tree, and everyone tried to show how their new initiative was related to it. This required its own kind of work—dealing with scammers who were trying to cash in by saying their program or project was culture related when it really was not.

More importantly, when interest in culture waned and it was no longer the flavor du jour, there was a temptation to try to hang culture on other Christmas trees—irregular warfare, security cooperation, information operations, great power competition, etc. While understandable from a budgetary standpoint (you have to show that you are contributing to whatever buzzword is getting the dollars), this may have contributed to the perception that culture was only valuable for a small subset of missions rather than across the range of military operations.

## *Tyranny of Metrics*

In the first few years of this cycle of DOD's interest in culture, there was a sense of urgency about developing culture-related capabilities and little pressure to demonstrate that we were doing the right things the right way and having the desired effects. At an early conference, one retired Marine colonel pointed out the absurdity of using traditional DOD assessment strategies when dealing with culture. He asked if the intent was for us to train one battalion, not train another, and then measure success by body count. Yet, over time, culture centers were called on to demonstrate effectiveness and efficiency, to justify their budgets and, at times, defend their existence. This pressure increased as the DOD's attention swung from operations in Iraq and Afghanistan to other mission types. Culture still was seen by many as something needed only for counterinsurgency operations and relatively few saw the futility of trying to conduct information operations or understand adversaries—large or small—without military personnel who could understand culture.

If the challenge had been to provide robust assessment

in the scientific sense, it would have been an adjustment for most culture centers, but a feasible one. Unfortunately, DOD consumes assessment almost entirely in terms of quantified information and thereby masks subjectivity, bias, and validity issues by turning information into numbers. The process of quantification also means that programs often are assessed on the basis of things that can be easily counted rather than on information that speaks to quality and effect. For example, when assessing training programs, throughput—the number of people trained—is easily turned into the sort of "metric" that fits nicely into the DOD's comfort zone but tells you almost nothing about the quality and usefulness of the program. Information from peer reviews of curricula (to assess quality) and post-deployment interviews (to assess utility and effect) is much more useful in an effort to truly assess the value of a program, but it is less easily turned into numbers that higher levels of the organization are willing to consume. For the Marine Corps' culture center, the answer was to create a dual track approach to assessment. We produced the quantified information necessary to feed the beast, but we also developed fairly robust approaches to assessing the quality and effectiveness of our programs. It still was not enough to keep our center from being closed.

The drive for quantification showed up in other areas as well. During the last decade and a half, DOD has spent countless hours and a truly astonishing amount of money attempting to come up with scales against which to measure the regional and cultural expertise developed by military personnel. These scales sometimes were developed by people with little understanding of how knowledge and expertise develop or the connection between knowledge and the ability to apply it effectively. One scale described its expert category as "able to pass for a native," and there were many similarly ridiculous criteria. The scales also all required that the capacity of individuals be measured, something on which the Services were reluctant to spend time or money. Most of these efforts collapsed under their own weight, and we were able to ignore them, but some vestiges of them did find their way into policies, which if followed as written would have left the Services trying to sort individual capabilities of military personnel into categories such as "fully proficient" or "expert" with poorly defined criteria and no additional resources.

Assessment is one area where we should have paid more attention early on. If the various Service culture centers had worked with the DOD-level policy offices to establish a reasonable compromise between quantified information and data that could speak to quality and effect, we might have been able to shape the assessment demands a bit more. That, in turn, would have made it easier to defend culture programs as the DOD's attention waned.

## Experts Inside and Outside the Organization

Across the DOD, we had challenges throughout this cycle with the way experts were conceptualized, hired, and maintained. Below, I briefly touch on some of the core lessons learned with regard to bringing experts on board and with trying to "outsource" expertise to academia.

### *Getting and Keeping Subject Matter Experts*

Across all the Services, recruiting, vetting, hiring, and keeping subject matter expertise was more challenging than I would have believed possible. Some organizations who were able to use the government civilian plan for faculty had more luck, but those of us who had to use contracts or the normal, general schedule (GS) government hiring system really struggled. Some hiring and contracting processes created unrealistic expectations for the depth and scope that can be covered by one expert. For example, finding one person who has regional or culture-specific expertise and also knows enough about the military to render the material relevant and accessible, understands adult learning and formal curriculum development, is experienced with in-person and online instruction, and knows how to design and conduct learning assessments is like winning the lottery. And yet, many contracts and hiring announcements were written in just that way. We could have done more to help culture programs clearly identify the capabilities needed and create staffing models that were more realistic.

Likewise, position descriptions and contracts often were imprecise about the qualifications needed. The vagueness was intended to provide organizations with some flexibility, but the end result was far too many mismatches in expertise. For example, a person who grew up in urban Kabul in the 1980s was not necessarily the right person to provide expertise to Marines dealing with shepherds in Helmand Province in 2010. I noticed this problem most often in trying to hire scientific personnel. Typically, those doing the hiring were not clear on the varied capabilities brought by different degree levels and would question why we wanted a PhD rather than somebody with a bachelor's and military experience. In fact, they sometimes did not even buy into the idea that a candidate's discipline mattered. We frequently were seeking someone like a PhD cultural geographer or cultural anthropologist but were offered somebody with a master's in business or regional studies. I am not sure what we could have done to improve this situation without fundamentally altering how contracting and civilian hiring is done in the DOD, but it was a persistent challenge.

The whole concept of subject matter experts (SMEs) in the DOD took some getting used to. DOD thinks of experts as people who are useful for topical knowledge that is already in their heads rather than people with the skills to create new knowledge. This creates three challenges. First, if you want the person to work with current information, you must build in time for them to maintain their knowledge. That may mean time to read, write, publish, and keep in touch with colleagues and support for events like conferences and symposia. Creating that time is difficult on the government civilian side and nearly impossible with contractors. Second, it can be very challenging to craft a contract or hiring process when you are seeking experts, particularly in the sciences, for their ability to find things out and develop new knowledge rather than for topical expertise. The focus on topical expertise is built into these processes and extra time (and patience) was necessary to work with the hiring system or contracting process to have any hope of getting the experts we needed. Third, at the current time, the DOD tends to treat experts as disposable commodities. The way most contracted and civil service jobs are constructed makes it very difficult for experts to keep knowledge and skills current, which can lead to them being let go or moved into different positions so that the organization can get somebody with more current knowledge. Also, if they are seen as troublesome (e.g., pushing for more robust or more accurate approaches), they may be edged out of an organization and replaced with somebody more compliant but potentially less qualified. This, in turn, yields situations that are not good in the long term:

- When an expert has a negative experience with a military organization or is let go, news is likely to spread in their professional community, making it harder to recruit highly qualified experts in the same field.
- When military personnel interact with an expert whose knowledge has been allowed to grow stale or who is not truly competent, they tend to remember. All too frequently, we saw military personnel take a negative experience with one expert and use it as a reason to dismiss advice from all experts in that field. In some cases, the experience was used as a rationale for dismissing the need for cultural knowledge and the organizations that develop it.
- Both of those outcomes have negative consequences not only for the individuals involved

but also for the sustainability of culture capabilities as a whole.

As a last point on subject matter experts, it is necessary to point out that the DOD as an organization has a relatively uncritical approach to expertise.[9] This is especially true in situations where the issue at hand is not something the DOD has spent a lot of time on previously and for which it has few knowledgeable individuals. Military organizations also have an understandable tendency to do the best they can with the smart folks they have on hand. If your expertise is in psychology, but your boss wants to know about neuroscience, you do your best to learn a few things on neuroscience and, the next thing you know, you are considered the organization's neuroscience SME. For us, this pattern manifested in a few fairly predictable ways:

- *The buzzword bandits*: these often are smart individuals who have become adept at reading the buzzword winds and tailoring their presentations of self to changes. So, they were counterinsurgency experts, then culture experts, then resilience experts, and so on. They may have an educational background that allows them to speak to each issue to some degree, but it is unlikely they could truly have expertise in so many things in such a short period of time.
- *The pet experts*: over time, leaders develop trust relationships with particular experts. This can be very useful when the expert has the confidence to know their limits but is dangerous when they behave in unethical or obstruction-

---

[9] This is a generalization. The DOD has plenty of smart, critical thinkers who are knowledgeable enough to make good decisions about expertise. However, they are not always in the right place at the right time and the organization as a whole suffers as a result.

ist ways. The ideal trusted expert is one who is happy to be clear about when they are not the right person and make a connection to somebody with more applicable expertise. Less than ideal pet experts use their special access to decision makers to restrict what the leader sees and hears, ensuring that they always appear to be the one with the answer.

- *True-but-trapped experts*: as mentioned above, sometimes a military organization has to make do with what it has on hand. This can sometimes result in the wrong person being in charge of an effort, even when they willingly admit to being the wrong person. If the trapped expert is not positioned to push back against unrealistic expectations, this can create serious problems with the organization's ability to get the knowledge it needs.

## *Break Glass in Case of Insurgency*
OUTSOURCING EXPERTISE

Getting experts into military organizations was hard. Keeping them and using them appropriately was even harder and maintaining a stable of experts was expensive. So, it is understandable that some DOD officials wanted to believe they could outsource their expertise requirements. One common refrain, especially as interest in culture waned, was that when cultural knowledge was needed, the DOD would be able to "reach out to academia" for expertise. This occasionally worked well, but we usually watched military organizations run into predictable obstacles.

Academic experts from civilian universities were used to building student knowledge of concepts during a semester or quarter and sometimes had difficulty adjusting their course content and delivery to "one-and-done" training sessions or distance programs with no faculty contact. Sometimes they were just too astonished at military approaches to culture-related learning to adjust. For example, the idea that you would expect military personnel to learn something useful about culture in a one-hour training session or that you would not provide language training to all personnel seemed bizarre. This led some academics to assume that the DOD was not serious in its approach to culture, or that it was simply seeking to check the culture box. Some also were reluctant to work with the DOD, for political reasons, out of wariness based on the history of the department's involvement with academia, or because they were concerned it would conflict with their discipline's ethical guidelines. Military organizations rarely had people with enough understanding of academia to anticipate or mitigate these predictable obstacles

The biggest challenge in outsourcing to experts in civilian academia was their ability to render advanced concepts in accessible ways within the time and format constraints of military learning. Many of the external academics with whom we worked were used to teaching concepts in a progression of classes from introductory to advanced. They tended to think of classes or curriculum development for military personnel as introductory because the students had no prior coursework. Introductory concepts, designed for getting across basic ideas in a discipline, often bear only a passing resemblance to the more advanced concepts actually used in the field. Yet, military personnel use knowledge in operational contexts and need the advanced concepts. Teaching military audiences requires translating those advanced concepts into forms that are accessible to people who have not had introductory and intermediate classes—work that is complex and time consuming. The work also required that the academic have some knowledge of how the concepts might be used by military personnel, not only to make classes more interesting but also to anticipate problems in how they would intersect with existing knowledge. For example, several of us, newly hired out of academia, introduced the concept of formal and informal economies not realizing it would be interpreted as legitimate and illegitimate rather than the way those concepts are used in social science. We sometimes referred to the work of translating advanced concepts as *up-armoring* them.

As with many of the other lessons learned, working with experts from civilian academia was not a problem with an easy solution. However, had we been better able to communicate the problems up front to the leadership of culture programs, we could have smoothed the path for colleagues joining culture programs straight from academia and perhaps more easily countered the calls for outsourcing.

## Conclusion

Despite all of the issues presented here and despite waning interest across the DOD, we did manage to get some things done. In retrospect, we accomplished more than I would have thought possible when I was first looking at the labyrinthine bureaucracy against which we would be pushing. Even those of us who worked on programs that have been greatly diminished or closed have left behind resources—books, curricula, and ideas in the minds of future military leaders—that will outlast us. I am not trying to paint too rosy a picture. There is no question that the cycle of interest and disinterest has played itself out again and that the DOD will have to relearn these lessons in the coming years, most likely at the expense of junior personnel and those with whom they interact. Still, I think there was enough about this cycle that was different, particularly the ability to leave artifacts and lessons learned behind in formats that will be discoverable, that we may have had a slightly more enduring impact than was possible in the past. Also, while the long-term impact is in question, the impact of the programs while they were running, at least those in the Marine Corps, is not. We have more than a decade of routine assessments and more comprehensive assessment research providing evidence of positive effect and, of course, also evidence of areas where we could have improved.[10]

As a final note, I want to highlight one lesson learned from this latest cycle often overlooked as we focus on what went wrong or right. The lesson is that collaboration works. I do not mean working with people you agree with or like or everyone coming together with some perfectly shared sense of common purpose. I mean finding points of intersection or mutual advantage and working together for as long as it makes sense. In my collaborations across the broad culture network over the years, there have been arguments, unresolvable differences of opinion, and many periods of mutual annoyance, but only a very few situations where we could not get past those things and find ways to help each other. Collaboration worked well to influence discourse, shape programs and policies, and build curricula. It worked well when people of different backgrounds—military or academic—worked together within a center, among counterparts across Services, and when people positioned differently in the landscape agreed on a goal and each worked it from their particular vantage point.[11]

Our collaborations required a willingness to compromise and, at times, to share or not get credit for an accomplishment. Those behaviors are not common in academia and, de-

---

[10] See for example, a few of the reports from CAOCL's assessment efforts posted on a DOD public portal: Wendy Chambers and Basma Maki, *Overall CAOCL Survey II Findings: The Value and Use of Culture by Type of Deployment* (Quantico, VA: Translational Research Group, Center for Advanced Operational Culture Learning, Marine Corps University, 2013); Erika Tarzi, *Regional, Culture, and Language Familiarization Program Messaging* (Quantico, VA: Translational Research Group, Center for Advanced Operational Culture Learning, Marine Corps University, 2017); and Erika Tarzi, *Educating Marines: Reorienting Professional Military Education on the Target* (Quantico, VA: Translational Research Group, Center for Advanced Operational Culture Learning, Marine Corps University, 2018).

[11] Kerry Fosher and Eric Gauldin, "Cultural Anthropological Practice in US Military Organizations," in *Oxford Research Encyclopedia of Anthropology*, ed. Mark Aldenderfer (London: Oxford University Press, 2021), https://doi.org/10.1093/acrefore/9780190854584.013.232.

spite protestations to the contrary, are not all that common in military organizations either. The benefits bear remembering for the next time when everything seems new and, as was the case in this cycle, the temptation to pretend to be a unique explorer discovering new territory is dangled in front of the next group to tackle these challenges. We, for the most part, managed to move past that temptation toward collaboration, and it was worth the effort.

CHAPTER NINE

# Alternative Perspectives
## Launching and Running the Marine Corps' Culture Center

*interviews with Jeffery Bearor and George Dallas*

## Introduction

This chapter provides alternative perspectives in two ways. First, it includes the reflections of practitioners rather than scholars. Second, it presents one perspective from the beginning of a culture program, the Marine Corps' center, and another from the standpoint of running the program for a decade and seeing it closed. The practitioner perspective is a vital part of understanding how culture programs developed and ran during this most recent phase of the Department of Defense's (DOD) interest in culture. The scholars who have contributed to this volume did not do their work alone and, for the most part, the organizational spaces and processes they worked within and sometimes sought to challenge were created and led by military personnel or civilian practitioners. We also wanted to take advantage of an unusual opportunity to capture interviews with the first and last leaders of one culture center, representing almost the entire arc of the center's existence.

The chapter is composed of two interviews. The first is with Jeffery Bearor who, at the time of the interview, was the assistant deputy commandant for Marine Corps Manpower and Reserve Affairs. While on active duty, then-colonel Bearor was deeply involved in the Marine Corps and DOD's early deliberations about culture-related capabilities. Immediately after his retirement in 2006, he became the first director of the Marine Corps' Center for Advanced Operational Culture Learning (CAOCL), which he ran for almost two years. The second interview is with George Dallas, who at the time of the interview was about to retire from his position as the director of the Center for Regional and Security Studies at Marine Corps University. While on active duty, then-colonel Dallas was the chief of staff for Marine Corps Combat Development Command (MCCDC), a vantage point from which he saw the development of many capabilities, including CAOCL. After his retirement in 2008, he became CAOCL's director and ran the organization until it was closed in 2020.

## Launching and Running the Marine Corps' Culture Center

There was about a year of lag between the two directors, during which CAOCL experienced significant organizational and personnel turmoil. It experimented with a series of leadership models, was briefly combined with the Center for Irregular Warfare, and endured the departure of many of its PhD subject matter experts and scholars who felt the environment had become either inhospitable or outright hostile. During this period, there were significant concerns across the DOD culture community that the organization would not survive. However, little documentation exists of this time so we must ask readers to make the leap between the leadership that launched CAOCL and its leadership for the remaining 12 years.

Readers may notice several areas where the interviews converge and diverge. For example, both directors transitioned from active duty service to the CAOCL director position within days. This provided both individuals with the kind of currency (in terms of relationships and working knowledge) needed to understand both Marine Corps requirements and the processes in place to meet them. Upon arriving at CAOCL, Bearor and Dallas both were confronted not only with the challenge of leading the Marine Corps' approach to culture training and education (outward focused) but also with the cultural friction that existed within CAOCL. Their experience confronting the kinds of clashes that often emerged when contractors, civilian academics, and active duty personnel (who often had very different ideas of what "right" looked like) worked together on various projects proved to be as difficult as the task of creating culture products. As one might expect, the ways in which they have reflected on and reconstrued these challenges was quite different. As George A. Kelly once said,

> A person can be witness to a tremendous parade of episodes and yet, if he fails to keep making something out of them . . . he gains little in the way of experience from having been around when they happened. It is not what happens around him that makes a man experienced; it is the successive construing and reconstruing of what happens, as it happens, that enriches the experiences of his life.[1]

The pages to follow offer moments of reflecting on and, at times, reinterpreting salient experiences associated with launching and running CAOCL. It is our hope that they will be of some value to those seeking to better understand the complexities associated with leading culture-related capabilities in the DOD.

## Interview with Jeffery Bearor, First Director of CAOCL
*Conducted by Kerry Fosher on 1 September 2020*
So, this would have been back in 2004, the march to Baghdad

---

[1] George A. Kelly, *A Theory of Personality: The Psychology of Personal Constructs* (New York: W. W. Norton, 1963), 73.

had already occurred. We'd been in Afghanistan for a little bit at various levels. And the fight going on in Iraq was morphing from the march to Baghdad into no kidding insurgency and a counterinsurgency campaign. My billet then was chief of staff at [Training and Education Command] TECOM and we were looking at the predeployment training program and what needed to morph from the old combined arms exercise and what became the integrated training exercise that prepared units to deploy to both Iraq and Afghanistan.[2] The focus now being mostly on Iraq.

General [James N.] Mattis came aboard to be the deputy commandant for [Combat Development and Integration] CD&I and [commanding general] CG MCCDC—brand new, promoted three star.[3] And he was very engaged in this process. In fact, he came to visit us within the first week of being the three star here at Quantico, and he kind of gave the CG, then Major General [Thomas S.] Tom Jones, a list of tasks. I was brought in on the conversations, and one of the things that General Mattis talked about was this transition from the march to Baghdad, "big war" piece down to a counterinsurgency, "small wars" piece.

We talked a lot about previous Marine Corps experience, both in Vietnam and back in the '20s and '30s when the Marine Corps wrote the *Small Wars Manual*, and he talked about some of the deficits in the training program, both at the individual level and on the unit level that he wanted us to get

---

[2] TECOM is the Marine Corps' organization that oversees all training and education efforts in the Service. Until 2020, it was run by a two-star general officer. It is now a three-star command.
[3] The Marine Corps has eight deputy commandants, three-star general officers responsible for different functional areas. CD&I is responsible for concept and capability development and determining requirements. It is broadly considered to be the most influential of the deputy commandant positions. MCCDC is the organizational structure supporting CD&I. The deputy commandant for CD&I is dual hatted as the commanding general of MCCDC.

at.[4] One of the things he talked about was this specific understanding of the people. He talked about, as we do in the *Small Wars Manual*, war among the people and the fact that the people are the focus.[5]

And of course, we're not the only ones to have figured that out. Certainly, you look at some of the insurgent campaigns, some of the ones that in particular were successful, Mao [Zedong], and what was going on in China. He knew that it was all about the people. So, his point was: what do we know about the people and what can we train our Marines to understand about how you influence people from a pretty different viewpoint?

Their culture's different. They had been living under Saddam Hussein for years and years and years. There were different factions. We didn't understand the religious factions, the political factions, and everything else. Nor did we understand—at the lance corporal, corporal, captain, and lieutenant level—what were the people's focus? What did we need to do in order to be successful, if you will, winning them over to our side and supporting them? Because that's what this was about. The various insurgencies were building up. What was going on in western Iraq, whether it was Sunni insurgents or Shi'i insurgents and all of that? And he thought that we needed a capability to get at that.

So, in fact, we weren't the only ones playing in the space. The other Services were as well. So, we stood up a series of working groups in order to pull in some experts and say: What do you need to do in order to provide the training and the expertise? Was there a language component? Is it just a cultural thing? What would the Marines need to know? And that's

---

[4] The *Small Wars Manual* is an iconic manual within the Marine Corps. Written in 1940 (and based on a 1935 manual on small wars operations), it was still considered important reading for Marines deploying to Iraq and Afghanistan in the 2000s.
[5] *Small Wars Manual*, NAVMC 2890 (Washington, DC: Government Printing Office, 1940).

how it got started. So, this would have been sort of the fall of 2004. And that was really the impetus. General Mattis coming in from his experience in both Iraq and Afghanistan and saying this is a deficit in our training program and we need to fill it out, which is why he came to TECOM.

*Fosher: So, with those groups that you brought together, what kinds of people did you pull into them?*

Well, we looked at our own staff and what was our biggest gap? Our biggest gap was cultural anthropologists and social scientists who understood the people piece. We looked at colleges and universities. We went to Naval Postgraduate School and looked for any experts that could come in. We reached out to the other Services and we found a pretty eclectic group, I think, of folks that come in and— You know the names as well as I do. It was pretty interesting. And we tapped into our own intel assets. Our intelligence assets also have some capability there across the Services.

Who was working in this space? Who could help advise us? What does a training program like that look like? We also then went back into our own history. We pulled out the *Small Wars Manual*. We pulled out training programs from Vietnam, to include language and culture programs, that we used it to pretty good effect in Vietnam for our advisor cadre and for the [Combined Action Program] CAP program, where we would match up a U.S. Marine Corps unit at the squad or platoon level with a South Vietnamese unit and put them together and then send them into the hinterland to work with the people in the villages, because that's what the North Vietnamese and Viet Cong were doing.[6] And so, we actually had some

---

[6] For more on the CAP program, see MSgt Ronald E. Hays, USMC (Ret), *Combined Action: U.S. Marines Fighting A Different War, August 1965 to September 1970* (Quantico, VA: History Division, an imprint of Marine Corps University Press, 2019).

background there. We just needed to resurrect some of those lessons and see how we can apply them into the environment we were in, particularly in Iraq.

So that's how we got that started. And there was quite a bit of push coming from [the Office of the Secretary of Defense] OSD in this same level, but I think actually the Services were ginning this up, particularly the Army and the Marine Corps, and [Special Operations Command] SOCOM to a certain extent as well.[7] And it was being ginned up that way. And OSD kind of glommed onto those efforts and provided some level of oversight.

*Fosher: That tracks with my understanding of how that went—that the Services energized OSD.*

And the good news was that OSD was willing to support, in fact, provided some pretty good resource capability in order to help, particularly the Army and the Marine Corps, get started down this track. And I thought there was a lot of good crosstalk going on between the Services, you know, Air War College down there in Montgomery, Alabama. They were also looking at this—it was pretty interesting—and it was a pretty far-flung group. We basically didn't turn away anybody nor their ideas because we knew we didn't have a monopoly on what needed to be done. About that same time, I gave [Center for Naval Analyses] CNA a task and to look across the entire training continuum, all of TECOM focused mostly on training—both individual and unit training—to see if we could identify any gaps in our formal learning centers where we might need to

---

[7] OSD is used here to refer to the broad organization of the headquarters-level staff in the Department of Defense. SOCOM is one of 11 unified combatant commands in the U.S. military. Each command has geographic or functional areas of responsibilities. SOCOM is a functional combatant command responsible for overseeing the special operations commands of the Services and commanding Joint special operations.

plug some of this in.[8] And they did a pretty good job, based on some guidance we'd given them, of looking across the curriculum of all the schools and coming back and saying, well, if this is your problem, these are your gaps.

About the same time, I think it might have been in early 2005, as we were getting through what the requirement was, we looked at what would be the broader requirement, particularly for individual Marine training and unit training through the [Integrated Training Exercise] ITX integrated training exercise.[9] In addition to the "language and culture" pieces, we decided we needed to look at all facets and levels of training, to include entry-level training to figure out if we had gaps. How could we better prepare new Marines for the training they'd receive when they got to their units and entered the predeployment workups? I tapped CNA to do a quick study. They identified some potential gaps and we sent the study results to all. We called all the schoolhouse curriculum developers and operations officers to Quantico, gave them a week to "fix" their [program of instruction] POIs, reviewed the changes, and changed the training programs to better align [entry-level training] ELT/[military occupational specialty] MOS training to help fill the training gaps.[10] That reduced the burden on units going through their [predeployment training program] PTP workups.[11]

As you can imagine, the intel school's curricula got quite

---

[8] CNA refers to the Center for Naval Analyses. The acronym is now used as the name for the broader nonprofit organization that houses the Center for Naval Analyses.

[9] In the Marine Corps, an ITX is a live exercise typically run as part of predeployment preparation.

[10] In the Marine Corps, ELT spans a broad range of time and settings from recruiting through recruit training (boot camp) or officer candidate training, several other stages depending on whether the Marine is enlisted or an officer, and typically concludes with MOS training where Marines learn their occupational specialty, such as infantry, intelligence, or artillery assigned during entry-level training.

[11] PTP includes all training a unit conducts in preparation for deployment.

a bit of an update. But even schools of infantry, we looked very hard there. We even looked at the recruit training for gaps in that. So, this was a pretty good-size effort and it actually worked out pretty well. We made some pretty substantive changes to the training programs so that we were starting to integrate.

Now, what do we need to do at the entry level? Training and MOS training schools? And then how was that going to bleed into unit training and what were the plus-ups there? It was only later that we kind of circled back to [professional military education] PME to see what other things should we be doing at places like Expeditionary Warfare School for captains, Command and Staff College, in particular, and even at the staff [noncommissioned officer] NCO academies— Was there a place there?[12]

And so, in the end state, what we got was based on those initial conversations with General Mattis. I think over a period of just about six months, we looked across the entire training and education continuum and we identified a gap. We had no capability inside TECOM in order to coordinate all that, nor to bring in the subject matter experts that we could then lend out, if you will, to the schools and to the training programs. And that's basically how we came up with the idea of, "Okay, we need a center."

*Fosher: I want to draw together two things that you just mentioned. One was the conversations across the Services and the other being the PME versus training aspect. I have a memory from very early on, just as I was beginning to get a sense of what the different Services were doing, of thinking there might be efficiencies to be found across*

---

[12] PME typically refers to the formal educational (versus training) programs. These are run by different schoolhouses and may involve resident and/or distance learning. In the Marine Corps, an NCO is an enlisted Marine in pay grades E4 (corporal) and E5 (sergeant). A staff NCO is an enlisted Marine in pay grades E6 (staff sergeant) through E9 (master gunnery sergeant or sergeant major).

*the Services. That the Air Force had the greater luxury to focus on PME, whereas the Marine Corps and the Army had to be focused more on training. Was that accurate and is that something that you were thinking about or that other people were thinking about? How to draw on those differences?*

So, it's interesting. You probably have a copy—I certainly do—of the initial draft brief that talked about the sweep of the program.[13] And in fact, that was the first time where we talked about how this cultural training and education program had several different pieces. A lot of it was focused on individual and unit training, predeployment. Okay, where are you going to go? What are you going to be doing? Again, we focused on Iraq. What part of Iraq? Where are you going to be? And what was the mission going to be of your unit? Because an infantry unit might have a different vision or different mission than a logistics unit, although everybody was going to be outside the wire in with the population. What did an infantry squad leader need to know that perhaps a logistics section chief didn't need to know? And again, trying to figure all that out.

One of the other things though that General Mattis talked about and that we incorporated in the initial vision of this thing was what are we going to do forever? In other words, his vision of this was that these are capabilities that we should have never let go of, as we always do. Okay, so into the '20s and '30s, the Marine Corps had been in small wars all up and down Central America and into the Caribbean.

We learned a ton of lessons. We actually put them into a manual so that we would have that. And then, of course, we had big war. And now we're an amphibious force—500,000 Marines, six divisions, fighting in the wars in the Pacific, the Western Pacific, South Pacific—and small wars got put on the

---

[13] This brief was developed in 2004 and used in updated forms through 2006. A copy was not available from the interviewee at the time of publication.

sideline. Then, of course, our next war was another pretty big war in Korea and we didn't need those capabilities.

But then our next war was not. It was small wars. It was war among the people. And so, we dusted off some of those ideas and tried to incorporate them into what we were doing in Vietnam. And of course, soon as we got to Vietnam or out of Vietnam—this is when I first came in the Marine Corps—nobody wanted to talk about small wars at all. We forgot the lessons. And now we're focused on northern Europe, particularly north Norway, the Baltics. How was the Marine Corps going to play in the big war, the coming big war against the Russians? We knew we weren't going to be in the Fulda Gap because we didn't have the capability.[14] So where were we going to play?

So, again, we forgot all those things. One of General Mattis' views was that this is something that Marines should do forever. You should never lose that capability. So that's how we came up with the [Regional Culture and Language Familiarization] RCLF program, where Marines would be assigned a focus so that we would always have some number of—I wouldn't call them experts—I'll call them literate Marines who would understand the general principles of how you operate in small wars among the people and would have some understanding of some region of the world.[15] Not to a huge, great depth. But the theory was that if you had enough of these Marines populated throughout the Marine Corps, every unit would then have some understanding and some depth of

---

[14] *Fulda Gap* refers to the lowland corridor running southwest from the German state of Thuringia to Frankfurt am Main. After World War II, it was identified by Western strategists as a possible route for a Soviet invasion of the American occupation zone from the eastern sector occupied by the Soviet Union.

[15] RCLF was the Marine Corps' career-long, distance learning program for culture and language. It was run by the Center for Advanced Operational Culture Learning until 2020, when it transitioned to the Center for Regional and Security Studies. As of early 2021, the program has been defunded, but it is expected to continue running until the content becomes outdated.

knowledge. Not a great depth, but at least they'd know. Yes, I know where India is. I can show it on a map. I understand the religious complexities of the country and the demographic complexities of the country. I know that they've been at war with the Pakistanis for 65 years and this is why. And you can start to now explain that and then, even more, know where to go to get the answers.

So, this was the other part that, quite frankly, after some of the efforts in Iraq and Afghanistan started to wind down, I think we lost sight of, because even though I think it continued to be a program, it was nobody's focus of interest anymore. Part of the reason is, you know, once again, because it's difficult and we've been doing it for 14 or 15 years now. People want to forget that piece and now move on to something else.

So, we're in a bit of an inflection point right now in that respect. Are we going to forget all the lessons that we've learned over the last 15 years—all of them hard won with blood—to get ready for the next big thing? Are we going to continue to incorporate those lessons as we go forward? Because if you're in a Marine littoral regiment and you're operating in the Philippines or Indonesia or with the Malays, you're going to have to understand the people you're operating with. I mean, even when we go to Australia, we find that there's differences, much less operating with Koreans and Japanese and the folks in Southeast Asia who are sometimes our allies and sometimes not. If you're going to operate with those folks in their country, you've got to understand them. So, we may be throwing out some baby and bathwater here.

*Fosher: I think the way we started phrasing it after a while was "partners, populations and the adversary." I want to step out of this history for a little bit and ask you to reflect on one or two things you found unexpectedly intellectually challenging or interesting during your time working on this.*

Well, a couple of things. One of the things that struck me in a good way was that Marines, once you started talking to them about this tool in the toolbox, they got it. Junior Marines understood that. I need to understand that if I see this scene on the street, what it means, you know, is this normal? Is this a normal street scene? Is this an abnormal street scene? And in fact, we bled that into a lot of the training. When you go to the infantry training lanes and the villages that we built, we tried to be able to paint that so that Marines would understand what looks normal and what looks abnormal so they can start to make some decisions.

And this was incorporated in a couple other things like Combat Hunter and some of the other things that we did on the training end.[16] Once you started explaining how this cultural understanding could affect the mission and would be a tool that they could use to understand what they were seeing, particularly, walking patrols in Ramadi or Fallujah—or pick a place in Iraq—or even some of the places in Helmand in Afghanistan. They started to understand. They went, "Oh, okay, I understand why this is important to me. It's important for me to understand the street scene with mom, dad, you know, the vendor, the store, open or not, as it is to be able to see if somebody's carrying an AK-47 or not, because it's going to give me information."

Marines glommed on to that pretty fast. It didn't take them long to understand that this was good information for them to have and helps them understand what they see once they got outside the [forward operating bases] FOBs. I was very, very encouraged by that. Even junior Marines understood that they needed to be able to see and understand the battlefield that they were on. And this included a hard look at local customs and village/town daily routines—what was

---

[16] Combat Hunter is a Marine Corps training program focused on developing advanced skills in observation, profiling, tracking, and questioning and also includes material on policing in a combat environment.

"normal" what was "abnormal"—the people and the scenes, because that would help them accomplish their mission. So that was a good thing.

The other side of that coin was that we were trying to jam more training into a very narrow space in time from the training programs to the deployment schedules that drove them. That meant that some things got short shrift. And a lot of times it was this piece, even though Marines understood that it was important to them. In a training schedule, it's got 150 days of training crammed into 100. What gets cut? And sometimes I was a little bit taken aback by it. Got it. You know, you have to be able to move, shoot, and communicate, do all those things that Marine units do. But sometimes the thing that got cut was this very thing that would allow them to understand what they were seeing.

That's one of the reasons why we wanted to incorporate some of this training into Combat Hunter, which was happening at the [School of Infantry] SOI.[17] So, we controlled that at TECOM and those Marines coming out of that training would have at least some common baseline understanding. And again, that way those Marines utilizing the training programs and the training information that we would provide through CAOCL, a lot of them would do that training on their own.[18] So, again, there's always that balancing act. We know we need more and better, but we get constrained by time.

*Fosher: There was, for what it's worth, very good partnering between*

---

[17] SOI has East and West Coast locations. The school provides classroom, hands-on, and live-fire training to develop combat skills. The school has programs for both infantry and noninfantry Marines that differ in content and duration.

[18] From 2006 to 2020, CAOCL was the Marine Corps' center for training, education, institutionalization, and other support related to the learning domains of language, regional knowledge, and cultural knowledge.

*Combat Hunter and CAOCL on the culture-general front.*[19] We consulted on some of the material they created on observation skills. And later, we incorporated some of their material into the culture-general content. That's the bulk of the content that we're leaving behind that I think will have real staying power, because it won't change as frequently as the culture specific material.

And again, this is that notion of why cultural understanding is important. There is never enough time during the predeployment training phase to accomplish everything the unit/ Marines need to do. Could we determine what "culture 101" training could be "off-loaded" back to the schoolhouses so that the units could focus on that language/culture training particular to their mission and their deployment geolocation? Because whether you go into Iraq or Afghanistan or to the southern Pacific or Africa or South America—wherever you're deploying. Having that basic understanding from prior education helps, even if it's just that there's a requirement to know the people and the culture you're going to be embedded in, whether it's with allies or adversaries. If that understanding is already there, then you can go get the specific information whether in predeployment training or during deployment.

So, making that part of preparation for battle, preparation for deployment sets us up to be more successful in future endeavors. Getting the specific information is relatively simple. There are plenty of experts out there that can help you with that.

*Fosher: You covered some of this previously, but I want to get at it a little bit more specifically. What problems could the loss of the cultural capability present for the Marine Corps or DOD going forward,*

---

[19] *Culture-general* refers to an element of culture learning focused on concepts and skills that can be employed in many different places. It complements culture-specific knowledge, which is focused on the details of one particular group or area.

*particularly given the way that the security environment is being cast right now?*

Well, it's interesting. The current *Commandant's Planning Guidance*—I'm a fan.[20] I think we have needed to kind of change our approach for a while now. We do have a rising near peer competitor. Still, the chances of combat in the South China Sea are pretty small, particularly combat that would require hundreds, if not thousands, of Marines. Just like the chance for combat on the Korean Peninsula is likely pretty small, but still there. So obviously, you have to prepare for the most dangerous outcome. But the most likely outcome is small wars and it remains small wars. Whether those are our particular and peculiar to places where we still have Marines deployed like Iraq and Afghanistan and Syria and the Horn of Africa and other places—those things aren't going away.

And this capability promotes success in those environments, which are still with us. And as I said, we've got Marines deployed in combat today all around the world. They're not fighting the Chinese. They're fighting small wars among the people where this sort of information and this training is going to be vitally important. And again, history tells us that we'd like to forget these lessons as soon as soon as we're done with the fight. We saw that in Vietnam. We saw it after the '30s. So, I think there's risk. I think there's risk that comes with not keeping this particular tool ready to employ.

*Fosher: You mentioned that even junior Marines kind of could fairly easily wrap their heads around this early on. And that tracks with my experience, too, with teaching culture-general. I was frequently just giving people words for things that they already knew. They just didn't have a basket to put it in. What are a couple of other things that went right?*

---

[20] Gen David H. Berger, *Commandant's Planning Guidance: 38th Commandant of the Marine Corps* (Washington, DC: Headquarters Marine Corps, 2019).

I talked about the relationships with the other Services and with OSD—the folks that were working in this particular part of the training fight, if you will. It was a pretty synergistic group, the ability to exchange ideas freely to talk about, "Hey, here's my problem. What's your problem? How did you solve it?" There was a lot of crosstalk.

We hosted conferences and meetings where everybody was invited and everybody got their say. I thought the synergy was really good. We were getting a lot of support, for instance, from [Under Secretary of Defense for Personnel and Readiness] USD (P&R), particularly on the readiness front.[21] We were getting a lot of support from some of the leading lights the General [Robert H.] Scales, General Matisses of the world on how important this piece was. And I thought the synergy was pretty good, particularly when we were getting this thing up and running over the first couple of years—the 2004 to 2006 time frame when this thing really took off. That was really good news—the ability to share the information and to take the best of the lessons learned and apply them. Then there was some pretty good resourcing being put against this. The combat villages that were being built, the FOBs and everything else that we did in places like Twentynine Palms [east of Los Angeles], Fort Irwin [northeast of Los Angeles], at the training base in Indiana, down at Fort Polk [Louisiana], there were some good resources being put into setting the physical space so that it would support the training programs.

And then we were providing some of the cognitive tools so that Marines and soldiers in particular would get that proper predeployment training to make them more successful when they went forward. So, there was a lot of good synergy there. And, of course, there were a lot of resources available to put against it. I think we all knew early on that that resourcing

---

[21] In this context, USD (P&R) refers to the office in OSD rather than the individual. P&R covers a very large range of functions including education and training as well as readiness.

capability wouldn't last forever, but we were able to get a lot of stuff done very quickly.

Of course, a lot of those training capabilities still exist. We still have the largest urban training space in the world out at Twentynine Palms.[22] And we use it for a lot of stuff. We use it for high-end training. We use it for predeployment training still for the units going to Iraq and Afghanistan, maybe at a smaller level. And again, it's provided us some pretty good capability. So, the willingness of everybody to play well together and then share ideas. I thought it was really pretty good.

*Fosher: The last part of my chapter is about that same issue from the scientific side during that period of time when most people put aside their egos and just got on with it.*

Well, so that's another interesting piece. The people who have this information in many cases weren't people who would normally associate their particular capability and their particular skill set with supporting the military, with being engaged in a military training program. In fact, it's not what they did at all. We got quite a bit of support. Some people would say, "No, we don't want to help you out." But there were plenty of people who said, "Yes, we understand you have a deficit. We see how we can help. And we're going to and we're willing to do that." So, there was enough support where subject matter experts from both inside and outside the government, to include from academia, said, "Yes, I'm willing to support. I see how my skill set can help." And I think we were well served in that arena. There were unique and new "partnerships" that had to be developed between the military, academia, and certain segments of the U.S. population that were, in some cases, not "usual." Social scientists working with the military and foreign-born U.S. citizens to support filling warfighting training gaps? Not

---

[22] This reference is to the Marine Air Ground Task Force Training Command at Twentynine Palms rather than to the town.

the usual cast of characters, and pretty unique in many ways.

Regardless of what's going on today or not, the fact of the matter is that America has somebody from everywhere and those folks from everywhere that we tapped into, particularly from the Middle East or Afghanistan, they showed up and they applied their expertise, having grown up in the environment, to help us get this training right.

And I found that to be pretty, pretty humbling that folks from all over the world, when we asked them, said, "Yes, you know, this is our adopted country. We understand we have a skill set that might help you." And of course, concomitantly they were helping the folks that they left back in, in particular Iraq and Afghanistan. So that sort of support was very vital. And I thought it was pretty humbling that these folks would show up. Now, we had some turn us down, but we had enough folks step up that it made it, I thought, a pretty good training program.

*Fosher: Okay, last question for me, although you're welcome to go on with anything else you're interested in. What recommendations would you make to the people who in 5 years or 10 years or 15 years have to build this capability again?*

So, we're probably better at this than we used to be. My assumption is, not having been part of this for years, is that we have created the record. That the training programs are stashed somewhere. That we have some sort of warm base to start from and that we're not going to have to go search for training programs, supporting programs, constructs, if you will, from training like we did when we first started this. That we have created that record, that it exists somewhere, hopefully at the university and within TECOM, so that the next time we have to start something like this up, we don't start from nothing.

We didn't start from nothing. In 2004, we had the *Small Wars Manual*. We had a lot of things that we started dredging them up from training programs in Vietnam. We had people we could go to and say, "Do you remember going through your language and cultural training program before you deployed as an advisor to the Vietnamese Marines?" We had those people around and we could tap them, say, oh yeah, this is how that training program looked.

We went and we found curriculum. The cultural and language part of the curriculum was going to be different. But the construct was useful. "You need to know this. You need to know this. You need to know this. These are some training opportunities and programs you should deploy." And so, at least we had something, but it took quite a bit of time to pull that information together. So, going forward, we need to establish that warm base.

We need to know where the information is. We need to be able to pull that up and again, maybe take the curricular construct and then apply the new information against it and be able to roll out a training program much more quickly than we were able to do then. Although, we turned product pretty fast back then.

We know that requirements are going to shift over time as long as we take the lessons and put them somewhere that they're available. Again, knowing that next time this happens, wherever it is we go, that we're going to have to put together very similar training programs.

We have to know: How did we do it last time? And again, if we're better at that than we were out of Vietnam or even out of the small wars of the '20s and '30s, then we'll be able to ramp up much more quickly. The other thing is, we were very lucky that we had General Mattis to press us along. He kind of had already figured this out. He knew what the requirement was.

And he was very, very articulate in laying out his vision. He didn't tell us how to do it, but he said, here's what the state needs to be in. And he was very good at that. And he gave us quite a bit of time early on. "Okay, General, here's what we thought you said. Here's what we've learned. What do you think?" And he would actually give us some broader guidance and kind of kept us on track. And the entire time he was at CD&I, when he was the deputy commandant there, he gave us a lot of his time and effort, which I very much appreciated. That helped keep us on track.

So hopefully next time this happens, we will be able to find the materials from this time and draw on Marines with relevant experiences. It's interesting that 10 years ago, we were nearly 85 percent combat vets. Now, even today and in the year 2020, the number of combat vets is going down pretty fast. Almost every general officer over the last couple of years has combat experience. But it's really instructive how few of the current battalion commanders have any combat experience other than as platoon commanders, perhaps as company commanders. And so that goes away very fast.

The senior leaders that have lived this dream and were company commanders and maybe even battalion commanders who are now general officers, they're going to have to be the ones who, hopefully, recognize that this sort of capability that we've got to have, if nothing else, on the warm base, ready to heat back up if we need it.

You don't find visionaries like General Mattis everywhere. But there should be enough around who have this experience, who can help drive this thing. Because it's going to happen again. We all know that. It's just what's the cycle? What's the time cycle?

It was an interesting time. You know, I only formally did that job for two years, although I spent almost the last probably near year I was on active duty doing it while I was chief of staff. Why me? I sat there with General Jones and heard General Mattis talk about this. And so once General Mattis left, the CG, Tom Jones turned to me and said, "Okay, who ought to be doing this?" And we talked about that for a bit. And I said, "I'm going to have to do it because I, as chief of staff, could put my hands on all the levers." And so that became my part-time job. And then it became a full-time job when I retired. That's how it came to me.

It was interesting that at that point, as we looked across everything else that was going on in TECOM, you know, could the G3 [operations and plans] do it? Should we give it to the Marine Corps University? Where would it live? And because we needed to touch training—entry-level through PME all the way to unit training out of Twentynine Palms, predeployment training—he said, "Well, you're the only one who can put his hands on all those levers." So that's how it got to me as chief of staff, not that it would normally have been the chief of staff's role to do it. So, then I transitioned right into the job, which is interesting as well.

*Fosher: When I was looking over that older transcript [from CAOCL's Oral History Project], I didn't realize that you didn't have much of a pause at all between.*[23]

No, I retired on Friday and I went to work on Monday. So how did that happen? We put together the initial construct of what the center should look like. We initially started with two GS15s—one to be the director, one to be the deputy director, sort of subject matter experts—because we saw you're going to have to have somebody in charge and running the thing, running the business and somebody who really knows it.[24] When we were having that particular conversation going back to General Mattis saying, "Okay, this is what we're going to do, this was going to stand up, that's how much it's going

---

[23] Kristin Post, "Interview with Jeffery Bearor," Translational Research Group: Center for Advanced Operational Culture Oral History Project, 6 March 2013. Note: interview transcripts from this project, along with other Translational Research Group and CAOCL materials have been archived with the Marine Corps History Division, Quantico, VA.

[24] This refers to level 15 of the general schedule (GS) pay plan for federal civilian employees. GS15 is the highest level in the plan. The next highest level in civil service is within the Senior Executive Service.

to cost. This is the initial lay down. Do you think we missed anything?" And, I remember him saying, "Well, who's going to be in charge?" We said, "Well, we'll hire somebody." And he said, "No, you're going to be in charge."

That was a pretty quick two years as well, because at the same time that we were getting CAOCL up and running, toward the end of the first year, we were talking about the Small Wars Center.[25] How does that work and how are we going to take the other lessons learned, not just the cultural training lessons, and how are we going to roll those in? And so, I spent the second year putting together a small wars center as well, which then got transitioned over to MCCDC. So, it was a pretty quick two years and then I was out there out the door.

## Interview with George Dallas, Director of CAOCL from 2008 until Its Closure in 2020
*Conducted by Lauren Mackenzie on 20 August 2020*

*Mackenzie: Can you begin by describing how you got involved in culture efforts in the DOD?*

Well, I just think through positions and beliefs. You know, probably as an early field grade officer, you begin to realize the importance of relationships of people and, of course, understanding culture is the foundation of understanding people and understanding relationships and cultivating relationships. So, I think, as a field grade officer, you begin to get exposed to these things. You begin to observe and see how important it is to do in order to get things done.

The relationships that you establish now will pay dividends well into the future. And you see the requirement to establish those relationships, which is really an understanding of

---

[25] *Small Wars Center* refers to the Small Wars Center and Irregular Warfare Integration Division that is part of CD&I under MCCDC.

the people. And so, I got to see that. I got to see that firsthand again through my assignments, the responsibilities of those assignments. So, it became very clear fairly early on, kind of midcareer level, the importance of it.

So, I've always had that interest. I've had many failures at it. And many times, you kick yourself in the pants. If you'd only known, if you'd only taken the time to understand, you would have been that much more effective as well as much more efficient. So, through my various assignments, through failures, disappointments, and I had some successes as well. I was able to learn and understand the importance of all this.

And then that, of course, generated my interest in it. The other thing is for me, at least, I got to play in many, many different cultures on every continent. So, I really had a good understanding of the variance between cultures and how somebody can look at something and see something completely different than I am. You talk to those general culture skills, but nobody ever called them that.

But you see it. So, that's how I grew to understand the importance of culture. And then I had an opportunity, as I was getting ready to leave active duty, to look within the Marine Corps, within government. I had obviously many opportunities outside of government in the business world because I had a fairly good Rolodex, as they say. But none of the jobs that corporate America was offering. . . . I mean, they all had challenges and they all had various points of interest, but none of them were involved with people as much as the CAOCL job was.

And I like people and I like dealing with people, despite the frustrations of it. It's also incredibly rewarding. So, that's what drove me to accept the CAOCL job when they offered it.

*Mackenzie: Can you say a little bit about the beginning days of you taking on the position at CAOCL? Any particular challenges that stand out to you?*

Frankly, when I got to CAOCL, it wasn't really in great shape. There had been a lot of infighting. It had essentially been leaderless for more than a year. It was a relatively new organization, so it had all of the relatively new organization problems, organization-wide focus, mission priorities. And so, it was pretty, I'll say, aimless in its direction, and the early challenges were just to get your hands around it—kind of established lanes, priorities and those kind of things, just basic organizational skills and less content and material solutions. So, I spent most of my time initially just building the box in which CAOCL would operate.

*Mackenzie: So, what kind of knowledge do you feel you needed to develop as you progressed in your position as director of CAOCL?*

Well, again, because of previous assignments, I had a pretty good grasp on the business side of things. I knew how to run an organization. I knew how to set priorities. I knew the budget.

Probably the weakest area that I had from a business perspective was the understanding of contracts and the dos and don'ts of contracting. There are legal ramifications to that, so where we may have tended to be a little more seat of the pants, as you did things in the military, we had to be more careful. And so that was probably a weaker spot that I had. And we were primarily a contractor-supported organization, probably at the time, 70 percent, maybe 80 percent of the people that worked at CAOCL were contractors. So, from a business perspective, it was understanding contracting from a substantive perspective.

I had a lot of practical experience. I had a fairly good understanding of the dos and don'ts of many, many different cultures because of my immersion throughout my career within those cultures. But that was primarily superficial stuff. So, from a substantive position, I needed to understand, I would say the academic side of it. You know, the Marines tend to be very pragmatic and so understanding the academic underpinnings of the culture was important. Training and the concepts of training—I did that my whole career, not hard education and the knowledge of the differences between training and education was probably a little trickier to get your hands around.[26]

But so, I would say the academic underpinnings—understanding in terms of the concepts and things like that—of what you've learned from a practical application side of the house, those I think were probably the biggest things for us to learn. And we didn't have, I mean, we had a lot of great trainers and training is very natural for Marines. But the educational aspects were a lot harder for us to grasp.

*Mackenzie: What are you most proud of in terms of what you were able to accomplish as director?*

Well, I'd say without trying to come across as too proud, I think our program, the Marine Corps program over the years became kind of the DOD flagship. We were the ones that were on the leading edge. We were the ones that were codifying the concepts, integrating them into actionable programs, whether they're training or education. We were the ones that were pushing the envelope beyond your basic PowerPoint or your basic understanding. I think the thing is we were always trying to do better, always trying to grow, always reaching for higher standards. And I think we did that the best, even though all four Services had their own programs and all four Services developed their programs to meet their Service requirements and all four programs were different.

There were still a lot of areas that overlapped. And I'd like to think that we were kind of the go-to guys. I think we tended

---

[26] For an explanation of how education and training were perceived in the DOD culture community, see Appendix: Common Culture Program Lines of Effort.

to be the ones that led. So, that was something I'm very proud of. You know, if there are any publications out there within the DOD community, we were the ones that published. So, with the exception, I think, of the Air Force—they had culture guides and we had culture guides—we had culture-general books and we had organizational underpinnings that people used, and people still reference today.

I think one of our early successes, to go back a little bit, one of the things we were doing when I took over was quickly realize that we were just kind of afloat. We were just out there, and we were responding and reacting to any kind of stimulus. And so right or wrong, we made a decision to go with Paula Holmes-Eber and Barak Salmoni's book about the five dimensions.[27] And what that did was it gave us an anchor point, a point from which we could shift from and maneuver from.

Instead of being all over the map and bouncing back and forth and not having any real direction, this framework put a spot on the ground and we could shift using our tools, and we would shift onto the target that we needed to fight. I think at the time, we were chasing the bullseye, and this gave us a point to shoot at so that I think that was an early success.

That may not have been 100 percent right, but it wasn't 100 percent wrong. And so, it allowed us to develop more focused and better programs because we had a starting point. So that goes back to maybe a previous question about our successes. I think in the end, we had a very holistic program that covered training, covered education, covered research. It covered all that, covered the waterfront.

And it not just covered them, but it covered them in detail, and it covered them in reasonable, responsible, and relevant

---

[27] Barak A. Salmoni and Paula Holmes-Eber, *Operational Culture for the Warfighter: Principles and Applications* (Quantico, VA: Marine Corps University Press, 2008), 24–28. A revised edition of this title was published in 2011. This book contains the "five dimensions of operational culture" framework that was used to organize CAOCL's training and education materials for most of the years it existed.

programs. If it was training programs, if it was educational programs, or if it was the research conducted by the [Translational Research Group] TRG team, it was a success.[28] I would go as far as [saying] no other organization in the Marine Corps, probably even in DOD, had an organization like the Translational Research Group, and that was the group of social and behavioral scientists that were able to look deeper at questions for us to really get to that rigorous academic depth, as well as provide research to broader questions for the Marine Corps. And that organization, you know, belonged to CAOCL.

I think it brought a very unique dimension to our culture center that the other centers weren't able to replicate. So, I think we had a lot to be proud of. And of course, I'm probably most proud of the people. We had great and we have great people and it didn't matter if they were government or contractor. They cared. And they wanted to do right by Marines, which is our purpose, our purpose is to serve Marines, serve Marines in this broader cultural area.

And they cared. The people that we had enjoyed working with at CAOCL and even the guys in uniform who came grudgingly, saying "What the heck's CAOCL?" "Why am I here?" At the end of the day, they saw the value and they enjoyed working in the cultural arena. And because they saw the value, they sold that understanding. We had people who understood culture was not just a necessity at the time, but it was a way for Marines to achieve their mission, whatever it is, if it was "bang, bang, shoot them up" or if it was more of a security cooperation thing, it helped Marines to more effectively, more efficiently accomplish their mission. And I would

---

[28] TRG was part of CAOCL from 2010 to 2020. It was a group of social scientists who supported CAOCL's concept and curriculum development, ran its assessment platform, and conducted research on problem sets relevant to other Marine Corps organizations. See Kerry B. Fosher, "Implementing a Social Science Capability in a Marine Corps Organization," *Journal of Business Anthropology* 7, no. 1 (Spring 2018): 133–52, https://doi.org/10.22439/jba.v7i1.5495.

say, though there's no hard evidence, there's no number out there, that we did save lives. I believe we saved Marine lives. And I believe we saved the lives of people in the battlespace. So yeah, I think there's a lot to be proud of. A tremendous amount to be proud of.

*Mackenzie: Thank you for that story. Maybe now you could transition to something that you're a little bit less proud of. Can you talk about any mistakes that you made?*

Oh! Every day!

*Mackenzie: Well, maybe early on, they could serve as cautionary tales for others in the future?*

Yeah, a ton of mistakes. Yeah, I mean, I think we can look at a lot of different things. So, from 2008 to about 2012, I would say the whole culture community, but we specifically at CAOCL, really enjoyed the support of the Services and of the Marines. It was visceral to them. They understood it, the importance of culture.

So as [Operation Enduring Freedom] OEF and [Operation Iraqi Freedom] OIF began to diminish, we had problems.[29] Frankly, from 2012 on, a whole bunch of my life was about keeping culture alive within the Marine Corps. I mean, we were a target from 2012 on—cut us, eliminate us, whatever. So that was the environment that I was dealing in, principally a defensive fight. Because of that defensive fight, I think we missed some opportunities because I was trying to save us for next year instead of perhaps looking deeper and moving faster to deeper targets. But I was concerned that if we didn't

---

[29] OEF and OIF were the two major combat operations in Afghanistan and Iraq at the time being discussed.

attack the immediate target, we wouldn't be there to attack the deeper targets as they came closer. So, as I look at it, the fact that we were put on the defensive very early on caused us to not be as quick as we should or could have been on some of the deeper targets.

RCLF would be an example. You know, we were kind of putzin' with RCLF for a long, long time. And even with Service buy-in at about the same time. You know, it still took us a long time to get the program up. It took us a long time once it was up to make that program reasonable and rational to the students that were taking it. A lot of issues with regard to assessment and the validity and, of course, concerns about death by PowerPoint.

So, there was the RCLF program, which was and is a great concept and a great program, which is another thing we led the way on. I mean, if you think about that program; it's incredible. But it took us a long time to really get it to a point to where it is: considered value added and not a pain. And, of course, the more pain you caused, the less likely it was that it would be accepted.

We finally got over that hurdle. But that's probably an example where we could have moved faster, should have moved faster. I probably should have named the culture center after somebody important, you know, get that name associated to it. That brings a lot of weight. There is no doubt in my mind that if we had named this something like [the] Mattis Center, there's no way that the Marine Corps would have walked away from it. No chance. They just would not have done that. It would have been worth the few millions of dollars that we get to just keep the name there. So that was a mistake, not so formative in the sense of culture, but a business mistake.

Marines, especially Marines that are on the line units, they just want the answer. Just tell me what I need to know. I'll focus on my warfighting and I'll incorporate this. But don't make me work for it. That's the culture-specific information.[30] Here's your facts. Do it. Execute and so that's pretty much what they want. So, concepts and the skills, the culture, general concepts and skills, they always kind of ran in the background. And we probably should have moved faster and bringing them to the forefront, especially as OIF and OEF drew down.

I think a lot of that stuff is fairly intuitive, but once you put a name to it, you kind of remember it better. And so, the idea of calling out the culture-general skills and knowledge and those kinds of concepts, I think that was an important opportunity that came out eventually, but probably should have come out sooner than later. So, I think that's an opportunity missed.

If we could have made RCLF a little easier to use, a little more robust, a little more relevant in terms of technology and the user interfaces and things like that, it would have been better. I think, generally speaking, the content of RCLF is solid, was solid, and it's only gotten better. Culture-general ran in the background and, while it was there, unless you call it out, it's harder to see and people don't necessarily relate it to things.

And so, we probably could have moved forward a little bit in a much more aggressive way on culture-general. The problem with that was Marines on the line. They don't like concepts. They just want the answer. So, it's kind of a balancing act. I think we did pretty well in broadening the aperture to include more conceptual material and skills as the environment changed, as direct contact with local populations decreased.

I think we did a good job of trying to capture and integrate the broader applicability of understanding culture. So, I think

---

[30] CAOCL made a distinction between culture-general concepts and skills, which can be used anywhere, and culture-specific information, the details of the cultural patterns in a particular area or group.

we did a pretty good job there, understanding that my big focus at the time was trying to keep something out there.

I think moving us to [Education Command] EDCOM was an absolute mistake.[31] And I say that for a couple of reasons. We already were somewhat of a bastard child. We were put into EDCOM and we were the absolute bastard child. Without a doubt. Nobody cared at all. I think that was a major mistake when they reorganized the TECOM headquarters. People struggle to value education within the Marine Corps and then you bury us inside education and our capability got lost completely. So, I think that was a mistake that was made by TECOM with very little input from us.

Let's see what other mistakes that we made. I guess the probably the biggest thing, again, I would go back to is this idea that from 2012 on, everything was a fight. That forced, well, maybe didn't force me, but I naturally looked closer, brought my horizons in, looked closer than I perhaps should have and didn't have that longer term vision. But with that said, I still think we were pushing the envelope more than many of the other culture centers.

And I would say that we were not unique in the fact of being undervalued. Everybody was feeling pressure. And as a result, today, there really aren't many true culture programs left.

*Mackenzie: Along those lines, why do you think there's been a decline in interest in resources, culture programs?*

Well, I think we don't go bang. We don't explode. Something doesn't break. And those kinds of programs don't do well in the military and particularly in places like the Marine Corps. So, we don't go bang. Our effects are silent. Our effects are

---

[31] EDCOM and Marine Corps University are, for all intents and purposes, the same thing. CAOCL was part of the headquarters of Training and Education Command until 2012. When the headquarters was reorganized, CAOCL became part of the Marine Corps University.

hidden from view, but believe me, they're there. We were also kind of a niche thing in the sense that Congress really doesn't care because we're not big employers. We're not running assembly lines. We're not doing anything on a major scale. We're decimal dust, not even decimal dust, in the big picture of things. So, we had no real advocate to carry the weight forward. And again, it's not just the Marine Corps. It was all the Services. And so, frankly, DOD had absolutely no clue.

I would say the fact that the DOD culture community—or not community—but that organization had really had no idea what it was doing and because it had no idea what it was doing, went immediately to the lowest tactical level, which was to create things.[32] You had four cultural centers working at the tactical level, trying to create things. So, from a community perspective, the fact is that we had no headlights out there. DOD, the Joint staff, those higher headquarters elements were all focused down instead of out, and they were giving us no direction, no guidance, no top cover.

And so, each of the four Services just ran their own little programs. We just lacked advocacy. Again, if we had named it the Mattis Center, we may not have had a *substance* advocate, but we would have had a *name* advocate. And no one would walk away from that. The name matters. So, the fact was that we just had zero advocacy.

And when the grassroots advocacy went away because Marines weren't using it every day, didn't see its impact because they were no longer engaged, it just kind of died. And I think it died a slow death, just like the Army's program.

*Mackenzie: Just two more questions. What problems could the loss of DOD culture programs cause for military personnel, in your opinion?*

---

[32] The reference here is to the DOD-level culture organization—the Defense Language and National Security Education Office (DLNSEO)—rather than to the broad community of people involved in culture programs across DOD.

Well, this is what we found after every war—we just recreate it again one more time. We pulled the lessons learned from Vietnam and, whammo, they're all the same ones and we'll do that again next time because we don't have the foresight, because it doesn't go bang. Because it's hard to make a direct connection from understanding culture to impact. And so, people say, "Oh, we can accept risk here."[33]

Well, you can to a degree. But what happened in '03, in '04, in '05, we were so focused on the kinetic and we opened a can of whoop 'em, but then we started to lose because all of a sudden impact in Baghdad didn't matter. It was the people that mattered, and we were not prepared. And so, then we scrambled, and we came up with all these great ideas and they were good ideas, but they're good ideas in extreme situations are very ineffective, very inefficient, and in many cases just bad. And so, we'll run that risk again. And we'll try to muscle through it. The shame will be the loss of people, the loss of equipment, the loss of ground that occurs from when the next war starts to the point we realize that we're losing because the people play an important role. It isn't like World War II, where the people didn't matter so much. The people will always matter now. Civilians will have a large say in what goes on in the battlespace. And so, while this cycle seems to be the routine, the consequences are just going to be worse.

*Mackenzie: And then finally, thank you for that, what recommendations would you make to the people who have to start up culture programs next time?*

Name it after somebody famous!

You know, this is really hard to do. And if you're like America,

---

[33] Here, the term *take risk* refers to supporting establishment and headquarters decisions to accept risk of capability loss by cutting programs, personnel, or other resources.

when you're under pressure, we just throw money at it. And frankly, we threw a lot of money at it. We had budgets early on up until about 2014, maybe '15, that were in $20 millions of dollars. But that was because we were doing a lot of hip pocket stuff. For example, somebody wanted a language program, so we'd go spend $3 million on Rosetta Stone in a year only to find out that, less than one-tenth of 1 percent of the people that signed up for Rosetta Stone ever completed it. But we had a language program. And so, someone can stand in front of Congress or in front of somebody with lots of stars and say, "Look, we got a language program," without ever noting that it was completely ineffective and a complete waste of time and money. But we had it, and that's all it seems to matter.[34]

So, one of the things I would say—and this will be hard to do—is to have a deliberate approach to think through what's possible. We had so many Monday morning quarterbacks coming from higher headquarters that gave you a solution in search of a problem.[35] We were the tail that was always wagging the dog early on.

Now, again, those kinds of activities are a natural occurrence in crisis environments. And so, it would be nice if our senior leaders could see this stuff coming. They get indications and warnings of attack when those indications and warnings are starting to appear. That's when we should be taking the

---

[34] "Corps Provides Free Rosetta Stone Software," Marines.mil, 5 December 2008.
[35] Here, "solution in search of a problem" refers to the fact that, in the early years of the culture program, many individuals and companies were seeking to align themselves with culture efforts, or more specifically, the money that was flowing to those efforts. During that period, CAOCL often was asked to consider adopting a "solution" that did not seem to help solve any of its problems.

programs out of the archives and making them happen.[36] Actually, we should never archive them. But if they're going to be there, we need to not wait to dust them off until people are dying or we're getting mission loss because we are failing in understanding the people in the environment.

And so, a more thoughtful and deliberate approach and one that is less reactive, one that is perhaps designed by the experts and not by the generals sitting around the table with a bunch of good ideas or Congress or something like that. So, I guess what I would encourage people to do is to act early. If you see something coming, start designing it. And one of the favorite comments from some people was, "Oh, we'll just break glass and pull the [subject matter experts] SMEs out."[37] That isn't going to work. It hasn't worked yet. And there are quite a few problems with that approach. But if you are able to begin actions when the indications and warnings start to appear and you have a deliberate approach to things, then I think you can get an effective program that has direction, that is relevant from both an operator's perspective and from an information perspective, and avoid the hip pocket knee-jerk reactions. I think that would be very helpful.

I think people have to understand that language is really, really hard. And if you want people to understand languages, that has to happen starting in kindergarten. You know,

---

[36] The term *archives* as used here refers to the process, common in military organizations, of storing lessons learned about past conflicts. As Jeffery Bearor mentioned in his interview, in the early 2000s, the Marine Corps managed to find some lessons learned and curricula from the Vietnam era to use as it thought about building culture and language programs. In 2020, as CAOCL was closing, George Dallas ensured that CAOCL placed examples of material and lessons learned in COLL/5918 Center for Advanced Operation Cultural Learning, 2005–20, Archives Branch, Marine Corps History Division, Quantico, VA.

[37] It was common across the Services to refer to all topical and scientific experts as SMEs.

from a more strategic perspective, the programs that support language development and those that have language requirements, those processes have to be more flexible and able to respond to the indications of warnings. This will be not an inaccurate factual statement, but it's the theme that I'm trying to illustrate. When we went into Afghanistan in 2001, there was like one Pashto speaker and, eight years later, there was one Pashto speaker in the Marine Corps. Again, it's not the number. It's the fact that we couldn't adjust to the demands. And, you know, we need to be able to do that.

At the higher levels, we need to be able to do that. Not an easy thing to do when you think about [Defense Language Institute] DLI and the programming that they need in order to bring on a bunch of Pashto speakers at the last minute.[38] But there have to be mechanisms that allow that to occur. And we've got to stop the silliness. Because we're not multilingual like many other cultures are.

And so, languages are really, really hard. I think we've got to get past the idea that the answer is a technological solution. The answer is people. And so, I think we need to be bigger than that. But that doesn't necessarily sell well, because the big companies prefer to offer you a technological solution and they convince their congressmen and senators or some general that it's the right answer.

But, you know, often these decision makers are just not going to know, and they trust. Understanding people, understanding culture is a very, very difficult subject, and it takes time and it isn't just about not using your left hands or not showing your soles of your feet. It goes much deeper than that. It's about how they think. If we understand how they think,

---

[38] DLI is an Army school, but it provides in-depth, resident language training to personnel from all military Services, foreign military students, federal civilians, and some law enforcement personnel. The school also provides immersion experiences and some online and nonresident learning options.

then we're inside their heads. If we're inside their [observe–orient–decide–act] OODA loop, we win.[39]

And so that's the value of culture.

You also asked about culture clash . . .

I would tell you that one of the biggest things that you have to deal with, if you're starting a program or if you're just, you know, maintaining a program is the culture clash that occurs within the organization itself. Our organization had active duty military. It had retired military. It had academics, government, operators, contractors, and subject matter experts that were government civilians or military or contractors. There are a lot of things that go on within that mix that you have to deal with every day. Academics do not see the world the same way the operators see the world. The things that are important to the operator are less important to the academics and vice versa. And so, you have this stew that's going on every day trying to understand and balance and bring out the best flavors in each one without curbing their motivation and initiative and drive.

This internal culture clash is a real issue to contend with. It is real. Certainly, I found that the issue between an Arab and a Jew less problematic. We never had issues that way. Where we had issues was the mix within. The academics would see it one way or the SMEs would see it one way. The active duty guys would see it one way and none of those three ways were the same. And there's added value into all of that. It's just finding the right balance in the right mixture.

## Conclusion

The quality and longevity of CAOCL's contributions to the

---

[39] The cycle of observe–orient–decide–act (OODA) developed by U.S. Air Force Col John R. Boyd is now commonly used in the military and other sectors to refer to the decision cycle. See Ian T. Brown, *A New Conception of War: John Boyd, The U.S. Marines, and Maneuver Warfare* (Quantico, VA: Marine Corps University Press, 2018).

Marine Corps (and well beyond) can be attributed to a number of factors, but both directors' commitment to learning and openness to new ideas set the tone for their staff to experiment with novel approaches to culture training, education, and research. Examples of this kind of growth mindset can be found in several of the themes that emerged from the interviews. Both directors emphasized how *people* and relationships were at the heart of and the impetus for their culture center's efforts. Although both knew full well the potential for cultural differences to lead to conflict and misunderstanding, they chose to focus on the benefits of culture training, research, and education for those producing, delivering, and receiving the content. Further, rather than frame the challenges—particularly those associated with the friction from civilian academics, active duty Marines, and contractors working together—as an either/or, win/lose proposition, they viewed the cultural differences as *both* problems *and* opportunities to learn from those problems. They both sought to bring their staff together to *use* what was happening internally at CAOCL to think through the kind of resistance that culture training and education inevitably faced. The effort to continuously balance training and education, distance and in-residence instruction, and academic content with military relevance required sustained attention to professional growth. This was something both directors not only demanded of their staff but also of themselves. They adapted during the years on a number of levels in response to internal assessments and a variety of external forces. Their goal always was to ensure the quality of the culture content their staff provided to Marines was something to be proud of. The sheer quantity and quality of artifacts that can be found in the CAOCL Archive at Marine Corps University is evidence of that.

Like many good stories, talk about ending CAOCL often brought its directors back to memories of its beginning, with the words of General Mattis's wisdom echoing in their ears. George Dallas contemplated the possibilities for CAOCL if it had originally been named the Mattis Culture Center. "The name matters," he said. Although there have been many names that have mattered to CAOCL over the years, none did more to get it off the ground and keep it alive than Jeffery Bearor and George Dallas. This book, in general, and this chapter, in particular, is an attempt to remind us of the importance of dedicated individuals, working in coordination from a range of different vantage points, to achieving complex goals like building and sustaining cultural capabilities.

# CONCLUSION

## by Kerry B. Fosher, PhD, and Lauren Mackenzie, PhD

The chapters in this volume cover more than 15 years of insights from individuals in a variety of roles and different organizations across the U.S. military Services. Chapters ranged across a wide spectrum of the work and perspectives necessary to implement and sustain culture programs. The book began with Ben Connable's observations of the need for cultural capabilities in wartime Iraq and his subsequent work in the supporting establishment. It then moved into chapters by Lauren Mackenzie, Susan Steen, and Angelle Khachadoorian, which focused on the challenges and opportunities associated with teaching culture courses in the professional military education context. The chapters by Anna Simons and Brian Selmeski formed a bridge between teaching and programmatic concerns, showing how both of those considerations played out in the development of culture-related programs. Allison Abbe and Kerry Fosher's chapters completed the transition by focusing on the programmatic aspects of the work necessary to launch efforts and keep them running. The book closed with reflections from leaders of one culture center, Jeffery Bearor, who launched the Marine Corps' center, and George Dallas, who ran it for most of its existence until it was shut down.

Despite the scope covered in the book, some themes do emerge. Several chapters highlight the difficulty of balancing traditional academic work, teaching, research, and maintaining one's expertise, with the weight of administrative work necessary to set the conditions for programs to succeed. In conversations, many of the authors expressed concern that they had not struck this balance appropriately. They felt they had either focused so much on academic work that they did not fully understand the bureaucratic gears and levers that controlled their context or, conversely, that they had spent so much time focused on programmatic issues that their time with students was minimal and their own scholarly expertise grew stale.

Several chapters also highlight the importance of balancing short-term wins with progress toward long-term goals. The Department of Defense (DOD) as a whole tends to be an impatient institution, often seeking tangible signs of success long before projects or programs are truly mature enough to be assessed. Reflecting back on the last 15 years, this was an area of weakness for many culture programs, especially given DOD's preference for quantified measures of progress and success.

Whether explicitly or not, all chapters speak to the importance of those newly arrived in the DOD taking time to build awareness of their context. For some of the authors, that meant taking time to learn about military students and translating material to be more accessible and relevant for them. As several authors noted throughout the book, this often takes more time than anticipated and there is no easy checklist to follow. For others, it meant learning about the existing discourses and processes of an organization and working with them or around them to get things done. All of the authors, at some point in their careers, also had to adjust to a context in which preexisting notions of culture as the "squishy stuff" conflicted with the martial orientation of students and leaders, where attitudes toward expertise vacillated unpredictably between blind acceptance and dismissiveness, and where there were strong preferences toward certain kinds of solutions that rarely included long-term institutional change.

Despite a sometimes-challenging context, the fact that all the authors in this book have persisted in their efforts to work with the military in some form also suggests that the work is worthwhile. Each of the authors has their own reasons for staying. Some hoped to create lasting change in organizations, others were captivated by the interactions with students and colleagues. Some even combined these two by, as a colleague once said of Anna Simons's work, changing the military one major at a time.

Throughout this volume, we hope readers also noted the importance of collegial relationships. The personal and professional connections between the authors were developed over many years, across different roles and organizations, and through many debates and arguments. They have been one of the more rewarding aspects of working within DOD culture programs and were instrumental to how we got things done. Building consensus allowed us to present a united front across Services, build on, rather than reinvent approaches and content, and work on issues from our different standpoints. The ability of people to build effective working relationships despite differences, competition for increasingly scarce resources, and the constant deluge of work to be done is one of the greatest strengths of culture programs in this cycle and a lesson we hope can be of use to those who come after us. It is worth the time it takes to build and maintain these connections.

The book was developed during a period when culture programs were in decline. The boom-and-bust cycle of the U.S. military's interest in culture had played itself out again. Yet, the writing in these chapters suggests there is some reason to be optimistic that the capability will not disappear quite so completely this time. For example, the Air Force Culture and Language Center is still in operation, if with a somewhat altered scope and mission, and the Naval Postgraduate School and Marine Corps University still employ faculty focused on culture and intercultural communication. Even as the Marine Corps' culture center was being shut down, the 2020 Naval Education for Seapower Strategy called for attention to adversary culture—a much reduced scope for the value of culture but still a nod to its continuing salience.[1] Also, as several authors noted, cultural capabilities have been tucked away under other names or in reduced form.

It is possible that the remaining culture programs and capabilities will fall prey to the tendency of the DOD to gradually shift uncomfortable ideas back into business as usual. In this case, that would mean a slow slide back toward the concept of culture being subsumed under the regional expertise and language programs that existed prior to 2004. Even if that happens, this most recent cycle is more heavily and publicly documented than was the case after World War II and the Vietnam War, which should make it easier for people to find places to start rebuilding capability. Nine years ago, in her article devoted to culture-related lessons learned during the Vietnam era, Abbe noted: "By incorporating culture into doctrine and into strategic guidance, the Department of Defense has greatly improved the odds that the cultural training programs implemented in recent years will survive beyond the conflicts that prompted them."[2] The chapters throughout this volume have illustrated that "survival" can take on different forms in the DOD, and the best we can do at this critical junction is to build on what we have learned. We encourage other centers and initiatives to follow the example of the Marine Corps' culture center and create archives of their programs, policies,

---

[1] *Education for Seapower Strategy, 2020* (Washington, DC: Secretary of the Navy, 2020).
[2] Allison Abbe and Melissa Gouge, "Cultural Training for Military Personnel: Revisiting the Vietnam Era," *Military Review* 92, no. 4 (July/August 2012): 9–17.

and courses.[3] Between enduring programs and records, we are hopeful that there will be a sufficiency of pilot lights left on to prevent a cold start the next time the DOD recognizes that it needs a more robust set of cultural capabilities to execute its missions.

---

[3] Shortly before it was closed in 2020, the Marine Corps' Center for Advanced Operational Culture Learning archived many of its guiding policies, reference materials, program descriptions, and some course materials, along with lessons learned in COLL/5918 Center for Advanced Operation Cultural Learning, 2005–20, Archives Branch, Marine Corps History Division, Quantico, VA.

# APPENDIX

# Common Culture Program Lines of Effort

This appendix provides brief descriptions of types of lines of effort or functions that were commonly found in military culture programs. Some programs may have used different categories. For example, the Marine Corps' center sometimes used the term *deployment support* to encompass several of the categories listed below, such as deployed support, reachback, support materials, and training.

**Analysis:** most commonly found in intelligence organizations or units, analysis involves gathering open source and classified information, evaluating sources, synthesizing relevant information, and reporting it in a format appropriate for a particular audience or mission. During the early years of the period covered in this book, several intelligence organizations had branches or offices focused on some form of cultural analysis.

**Cultural advisors** (see deployed support).

**Databases:** in the early years of the most recent upswing in military interest in culture, there was great interest in building databases of cultural knowledge and many initiatives were funded. These efforts failed to yield useful outcomes as they were based on an outdated concept of culture as a type of static system that could be broken into discrete parts and cataloged.

**Deployed support:** deployed support efforts typically involve sending one or more subject matter experts along with a unit or headquarters to provide in-depth advising to military leaders. In most programs, the subject matter experts were chosen based on a combination of having lived in or had deep experience in the area of interest and familiarity with military missions. In some cases, the expert would embed with a unit or headquarters staff during their predeployment process to become more familiar with the expected mission and the individuals being supported.

**Education:** educational functions are carried out in-person and through distance learning programs, usually, although not always, aligned with military schools and universities. Educational efforts often are contrasted with training efforts. Education focuses on deeper knowledge and/or how to think and training emphasizes the knowledge and skills needed for an upcoming mission or assignment.

**Mapping:** as with databases, military organizations have a great interest in developing maps of different aspects of culture, often seeking to understand the patterns and movements of kinship/political groupings, such as tribes, or patterns associated with religion, resource use, etc. Although visually appealing to military audiences and of some use early in conflicts, mapping approaches struggled to capture salient aspects of culture in operationally useful ways due to the changing nature of culture and variations in individual behavior.

**Modeling and simulation:** many military organizations invest heavily in modeling and simulation. There was an early expectation that these technologies could help predict human behavior in conflict or disaster zones and potentially reduce the cost of training by creating simulated environments. Models do hold promise in anticipating the range of human behavior, although not for prediction at the current time. Efforts to develop simulations for training typically did not yield usable results because the design emphasized the computational and visual aspects and too little attention was paid to underlying scientific realities of culture and the culture-specific details of the group being simulated.

**Predeployment training** (see training).

**Reachback:** reachback capabilities provide deployed military personnel with the ability to contact subject matter experts and others with specific questions. The reachback staff then conduct any research necessary to answer the question and provide a response in a format and timeframe appropriate to the situation. Culture programs differed in how they approached reachback, with some creating dedicated staffs and others relying on their education and training personnel to create responses.

**Research:** in most cases, the research conducted within culture centers was focused on supporting some other effort, such as curriculum or material development. Therefore, it emphasized either the scientific underpinnings of culture-related content or the culture-specific details of a particular group. The research often was done by the regular staff of the center rather than by a dedicated staff. Some programs also used research to assess the quality and impact of their efforts.

**Simulation** (see modeling and simulation).

**Support materials:** almost all culture centers spent a significant amount of effort producing materials to support learning.

Examples include smart cards, guidebooks, maps, textbooks, videos, and podcasts. Initially, most centers focused on developing products that could be used during deployments as reminders of (or substitutes for) predeployment training. Later, development expanded to include products that supported educational efforts.

**Training:** culture centers had training programs to provide personnel with specific information they would need for an upcoming deployment or assignment. Training was typically delivered in person, although the duration and content varied a great deal. In some cases, a unit would request a one-hour training session as their only preparation for navigating culture. In other cases, a commander might request several days of training, including specialized training for certain segments of the staff. In terms of content, training varied from short, basic regional overviews to in-depth classes on language, culture-specific details, and cross-cultural skills.

# SELECT ACRONYMS AND TERMS

All terms refer to U.S. entities.

| | |
|---|---|
| AAA | American Anthropological Association |
| AFCLC | Air Force Culture and Language Center |
| ARG | Amphibious Ready Group |
| ARI | Army Research Institute for the Behavioral and Social Sciences |
| ARO | Army Research Office |
| BIA/BIE | Bureau of Indian Affairs/Education |
| CAL | Center for Army Leadership |
| CAOCL | Marine Corps Center for Advanced Operational Culture Learning |
| CAP | Combined Action Program. This was a Vietnam War-era program that combined U.S. Marine Corps and South Vietnamese units for counterinsurgency operations |
| CCAF | Community College of the Air Force |
| CD&I | Marine Corps Combat Development and Integration |

| | |
|---|---|
| CG | commanding general |
| CEAUSSIC | AAA's Commission on the Engagement of Anthropology with the U.S. Security and Intelligence Communities |
| CJSOTF | Combined Joint Special Operations Task Force |
| CNA | Refers to the Center for Naval Analyses. The acronym is now used as the name for the broader nonprofit organization that houses the Center for Naval Analyses |
| COIN | counterinsurgency |
| Combat Hunter | A Marine Corps training program that focuses on developing advanced skills in observation, profiling, tracking, and questioning and also includes material on policing in a combat environment |
| CRSS | Center for Regional and Security Studies at Marine Corps University |
| culture-general | An element of culture learning focused on concepts and skills that can be employed in many different places. It complements culture-specific knowledge, which is focused on the details of one particular group or area |
| culture-specific | An element of culture learning focused on the details of the cultural patterns in a particular area or group. It complements culture-general learning, which focuses on concepts and skills that can be employed in many different places |
| DLIFLC | Defense Language Institute, Foreign Language Center |
| DLNSEO | Defense Language and National Security Education Office |
| DOD | Department of Defense |

| | |
|---|---|
| DODI | Department of Defense Instruction |
| EDCOM | Marine Corps Education Command |
| ELT | entry-level training |
| GWOT | Global War on Terrorism |
| HSCB | Human, Social, Cultural, and Behavioral. HSCB was a modeling program in the DOD's Directorate of Defense Research and Engineering |
| HTS | Army Human Terrain System |
| HTT | Army Human Terrain Team. HTTs were a deployed component of HTS |
| ITX | Integrated Training Exercise. In the Marine Corps, an ITX is a live exercise typically run as part of predeployment preparation |
| LIC | low-intensity conflict |
| LREC | Language, Regional Expertise, and Culture. Despite the "expertise" part of the acronym, LREC was used to refer to the full range of education and training related to language, regional knowledge, culture-specific knowledge, and culture-general concepts and skills |
| MCC | military culture center |
| MCCDC | Marine Corps Combat Development Command |
| MCU | Marine Corps University |
| MTT | Mobile Training Team |
| MOS | military occupational specialty |
| MURI | Multidisciplinary University Research Initiative. The MURI Program is a multi-Service DOD program that provides funds for science, technology, and engineering research and development within consortia of universities |
| NCO | noncommissioned officer |

| | |
|---|---|
| NPS | Naval Postgraduate School |
| OEF/OIF | Operation Enduring Freedom and Operation Iraqi Freedom. These were the two major combat operations in Afghanistan and Iraq at the time being discussed |
| ONA | Office of Net Assessment |
| OODA Loop | The cycle of "observe–orient–decide–act" developed by U.S. Air Force Colonel John R. Boyd. The term *OODA loop* is now commonly used in the military and other sectors to refer to the decision cycle |
| OSD | Office of the Secretary of Defense |
| RCLF | Regional Culture and Language Familiarization program. RCLF was the Marine Corps' career-long, distance learning program for culture and language. It was run by CAOCL until 2020, when it transitioned to the CRSS. As of early 2021, the program has been defunded, but it is expected to continue running until the content becomes outdated |
| ReARMM | Army Regionally Aligned Readiness and Modernization Model |
| PME | professional military education |
| PTP | Predeployment Training Program |
| PTSD | post-traumatic stress disorder |
| QEP | Quality Enhancement Plan. A QEP is a component of university accreditation under the Southern Association of Colleges and Schools, a regional accrediting body |
| RSEP | Regional Security Education Program. RSEP is an NPS program that provides focused seminars on regional and security topics both ashore and onboard ships |

| | |
|---|---|
| SACSCOC | Southern Association of Colleges and Schools Commission on Colleges. SACSCOC is a civilian body that accredits schools, including PME institutions within its region, to award degrees |
| SEAL | Navy Sea, Air, and Land team |
| SF | Special Forces |
| SME | subject matter expert |
| SOCOM | Special Operations Command |
| SOF | Special Operations Forces |
| SOI | Marine Corps School of Infantry |
| SO/LIC | Special Operations/Low-Intensity Conflict. This term can refer to a general category of operations or to the office within OSD |
| SOTF | Special Operations Task Force |
| TECOM | Marine Corps Training and Education Command |
| TRADOC | Army Training and Doctrine Command |
| TRG | Translational Research Group. TRG was part of CAOCL from 2010 to 2020. It was a group of social scientists who supported CAOCL's concept and curriculum development, ran its assessment platform, and conducted research on problem sets relevant to other Marine Corps organizations |
| USAF | U.S. Air Force |
| USAFA | U.S. Air Force Academy |
| USD (P&R) | Under Secretary of Defense for Personnel and Readiness |
| USMC | U.S. Marine Corps |
| USAJFKSWCS | U.S. Army's John F. Kennedy Special Warfare Center and School |

# BIBLIOGRAPHY

1st Marine Division. "Points from SASO Conference." Unpublished conference proceedings, 2002.
Abbe, Allison. *Building Cultural Capability for Full-Spectrum Operations*. Arlington, VA: U.S. Army Research Institute for the Behavioral and Social Sciences, 2008.
———. *Transfer and Generalizability of Foreign Language Learning*, Study Report 2008-06. Arlington, VA: U.S. Army Research Institute for the Behavioral and Social Sciences, 2009.
Abbe, Allison, David S. Geller, and Stacy L. Everett. *Measuring Cross-Cultural Competence in Soldiers and Cadets: A Comparison of Existing Instruments*. Arlington, VA: U.S. Army Research Institute for the Behavioral and Social Sciences, 2010.
Abbe, Allison, and Jessica A. Gallus. *The Socio-Cultural Context of Operations: Culture and Foreign-Language Learning for Company-Grade Officers*. Arlington, VA: U.S. Army Research Institute for the Behavioral and Social Sciences, 2012.
Abbe, Allison, Lisa Gulick, and Jeffrey Herman. *Cross-Cultural Competence in Army Leaders: A Conceptual and Empirical Foundation*, Study Report 2008-01. Arlington, VA: U.S. Army Research Institute for the Behavioral and Social Sciences, 2007.
Abbe, Allison, and Melissa Gouge. "Cultural Training for Military

Personnel: Revisiting the Vietnam Era." *Military Review* 92, no. 4 (2012): 9–17.

Agar, Michael H. *The Professional Stranger: An Informal Introduction to Ethnography*, 2d ed. Bingley, UK: Emerald, 1996.

"Air Force Culture and Language Center: Language Enabled Airman Program." AirUniversity.af.edu, accessed 22 March 2021.

"The Air Force Culture and Language Center's Voices of LEAP." YouTube, 6 February 2019, 3:20 min.

*Air University Fifth-Year Interim Report*, pt. V, *Quality Enhancement Plan Impact Report*. Maxwell Air Force Base, AL: Air University, 2015.

*Air University Quality Enhancement Plan, 2009–2014: "Cross-Culturally Competent Airmen."* Maxwell Air Force Base, AL: Air University, 2009.

Albro, Robert, George Marcus, Laura A McNamara, and Monica Schoch-Spana, eds., *Anthropologists in the Securityscape: Ethics, Practice, and Professional Identity*. Walnut Creek, CA: Left Coast Press, 2012.

Albro, Robert, James Peacock, Carolyn Fluehr-Lobban, Kerry Fosher, Laura McNamara, George Marcus, David Price, Laurie Rush, Jean Jackson, Monica Schoch-Spana, and Setha Low. *Final Report on the Army's Human Terrain System Proof of Concept Program*. Arlington, VA: American Anthropological Association (AAA) Commission on the Engagement of Anthropology with the US Security and Intelligence Communities, 2009.

Alison, Laurence, and Emily Alison. "Revenge versus Rapport: Interrogation, Terrorism, and Torture." *American Psychologist* 72, no. 3 (2017): 266–77. https://doi.org/10.1037/amp0000064.

Alrich, Amy A., Claudio C. Biltoc, Ashley-Louise N. Bybee, Lawrence B. Morton, Richard H. White, Robert A. Zirkle, Jessica L. Knight, and Joseph F. Adams. *The Infusion of Language, Regional, and Cultural Content into Military Education: Status Report*. Alexandria, VA: Institute for Defense Analyses, 2012.

American Anthropological Association. "Executive Board Statement on the Human Terrain Systems Project." Statement presented on 31 October 2007.

*Army Culture and Foreign Language Strategy*. Washington, DC: Department of the Army, 2009.

"AU's Cultural Education Efforts on Track, Growing." Maxwell Air Force Base, 16 December 2011.

Bacon, Lance M. "Commandant Looks to 'Disruptive Thinkers' to Fix Corps' Problems." *Marine Corps Times*, 7 August 2017.

Benedict, Ruth. *The Chrysanthemum and the Sword: Patterns of Japanese Culture*. Boston and New York: Houghton Mifflin, 1989; original printing 1946.

Berger, Gen David H. *Commandant's Planning Guidance: 38th Commandant of the Marine Corps*. Washington, DC: Headquarters Marine Corps, 2019.

Berry, John W., Ype H. Poortinga, Marshall H. Segall, and Pierre R. Dasen. *Cross-Cultural Psychology: Research and Applications*, 2d ed. Cambridge, UK: Cambridge University Press, 2002. https://doi.org/10.1017/CBO9780511974274.

Bourgeois, Jasmine. "AFCLC, Air University's First Virtual LREC Symposium Draws Thousands of People." *Air Force News*, 21 October 2020.

Brose, Christian. "The New Revolution in Military Affairs: War's Sci-Fi Future." *Foreign Affairs* 98, no. 4 (May/June 2019): 122–34.

Brown, Ian T. *A New Conception of War: John Boyd, the U.S. Marines, and Maneuver Warfare*. Quantico, VA: Marine Corps University Press, 2018.

Builder, Carl H. *The Masks of War: American Military Styles in Strategy and Analysis*. Baltimore, MD: Johns Hopkins University Press, 1989.

COLL/5918 Center for Advanced Operation Cultural Learning, 2005–20. Archives Branch, Marine Corps History Division, Quantico, VA.

Carter, Bradley L. "No 'Holidays from History': Adult Learning, Professional Military Education, and Teaching History." In *Military Culture and Education: Current Intersections of Academic and Military Cultures*. Edited by Douglas Higbee. Burlington, VT: Ashgate, 2010, 167–82.

Center for Advanced Operational Culture Learning. "Anonymized Interview from Longitudinal Assessment Project." In *Translational Research Group Report*, USMC IRB Protocol #USMC.2016.0005. Quantico, VA: Marine Corps University, 2016.

———. "Why Culture?" Quantico, VA: Marine Corps University, 2017. Video, 5:20 min.

Chambers, Wendy, and Basma Maki. *Overall CAOCL Survey II Findings: The Values and Use of Culture by Type of Deployment.* Quantico, VA: Translational Research Group Report, Center for Advanced Operational Culture Learning, Marine Corps University, 2013.

*CJCS Instruction 3126.01A, Language, Regional Expertise, and Culture (LREC) Capability Identification, Planning, and Sourcing.* Washington, DC: Chairman of the Joint Chiefs of Staff, 31 January 2013.

Connable, Ben. "All Our Eggs in a Broken Basket: How the Human Terrain System Is Undermining Sustained Cultural Competence." *Military Review* (March–April 2009): 57–64.

———. "Human Terrain System Is Dead, Long Live . . . What?" *Military Review* (January–February 2018): 24–33.

———. "Marines Are from Mars, Iraqis Are from Venus." *Small Wars Journal*, 30 May 2004.

———. *Military Intelligence Fusion for Complex Operations: A New Paradigm.* Santa Monica, CA: Rand, 2012.

"Corps Provides Free Rosetta Stone Software." Marines.mil, 5 December 2008.

"Culture: Culture Section." Defense Language and National Security Education Office, accessed 22 February 2021.

*Defense Primer: Information Operations.* Washington, DC: Congressional Research Service, 2020.

Deitchman, Seymour J. *The Best-Laid Schemes: A Tale of Social Research and Bureaucracy*, 2d ed. Quantico, VA: Marine Corps University Press, 2014; original printing 1976.

*Department of Defense Strategy for Operations in the Information Environment.* Washington, DC: Department of Defense, 2016.

Department of the Army. *Army Culture and Foreign Language Strategy.* Washington, DC: Department of the Army, 2009.

———. *Counterinsurgency*, Field Manual 3-24. Washington DC: Department of the Army, 2006.

Dweck, Carol S. *Mindset: The New Psychology of Success*. New York: Random House, 2006.

Earley, P. Christopher, and Soon Ang. *Cultural Intelligence: Individual Interactions across Cultures*. Stanford, CA: Stanford University Press, 2003.

Evans-Pritchard, E. E. *Witchcraft, Oracles, and Magic among the Azande*. Oxford, UK: Clarendon Press, 1976.

*Executive Order 13589 of November 9, 2011: Promoting Efficient Spending*. Washington, DC: Executive Office of the President, 15 November 2011.

Feickert, Andrew. *Army Security Force Assistance Brigades (SFABs)*. Washington, DC: Congressional Research Service, 2020.

Fisher, James L., James V. Koch, James T. Rogers, and Clifford L. Stanley. *Air University Review: February 2007–April 2007*. Venice, FL: James L. Fisher, 2007.

Fitzgerald, David. *Learning to Forget: US Army Counterinsurgency Doctrine and Practice from Vietnam to Iraq*. Stanford, CA: Stanford University Press, 2013.

Fosher, Kerry B. "Cautionary Tales from the US Department of Defense's Pursuit of Cultural Expertise." In *Cultural Awareness in the Military: Developments and Implications for Future Humanitarian Cooperation*. Edited by Robert Albro and Bill Ivey. London: Palgrave Macmillan, 2014, 15–29. https://doi.org/10.1057/9781137409423_2.

———. "Implementing a Social Science Capability in a Marine Corps Organization." *Journal of Business Anthropology* 7, no. 1 (Spring 2018): 133–52. https://doi.org/10.22439/jba.v7i1.5495.

———. "Pebbles in the Headwaters: Working within Military Intelligence." In *Practicing Military Anthropology: Beyond Expectations and Traditional Boundaries*. Edited by Robert A. Rubinstein, Kerry B. Fosher, and Clementine Fujimura. Sterling, VA: Kumarian Press, 2012, 83–100.

———. "Practice Note: Defense Discourses." *Anthropology News* 49, no. 8 (2008): 54–55.

———. "Review Essay: Anthropologists in Arms: The Ethics of Military Anthropology." *Journal of Military Ethics* 9, no. 2 (2010): 177–81. https://doi.org/10.1080/15027570.2010.491357.

———. "Yes, Both, Absolutely: A Personal and Professional Commentary on Anthropological Engagement with Military and Intelligence Organizations." In *Anthropology and Global Counterinsurgency*. Edited by John D. Kelly, Beatrice Jauregui, Sean T. Mitchell, and Jeremy Walton. Chicago, IL: University of Chicago Press, 2010, 261–71. https://doi.org/10.7208/9780226429953-018.

Fosher, Kerry, and Eric Gauldin. "Cultural Anthropological Practice in US Military Organizations." In *Oxford Research Encyclopedia of Anthropology*. Edited by Mark Aldenderfer. London: Oxford University Press, 2021. https://doi.org/10.1093/acrefore/9780190854584.013.232.

Fosher, Kerry, Lauren Mackenzie, Erika Tarzi, Kristin Post, and Eric Gauldin. *Culture General Guidebook for Military Professionals*. Quantico, VA: Center for Advanced Operational Culture Learning, Marine Corps University, 2017.

Fossum. Donna, Lawrence S. Painter, Valerie L. Williams, Allison Yezril, and Elaine M. Newton. *Discovery and Innovation: Federal Research and Development Activities in the Fifty States, District of Columbia, and Puerto Rico*. Santa Monica, CA: Rand, 2000, appendix B. https://doi.org/10.7249/MR1194.

Fox, Craig R., and Sim B. Sitkin. "Bridging the Divide between Behavioral Science and Policy." *Behavioral Science and Policy* 1, no. 1 (2015): 1–12.

Gelfand, Michele. *Final Report: Dynamic Models of the Effect of Culture on Collaboration and Negotiation*. Research Triangle Park, NC: Army Research Office, 2014.

Gezari, Vanessa. *The Tender Soldier: A True Story of War and Sacrifice*. New York: Simon and Schuster, 2013.

Goffman, Erving. *The Presentation of Self in Everyday Life*. Garden City, NY: Doubleday, 1959.

Green, Clifton. "Turnaround: The Untold Story of the Human Terrain System." *Joint Force Quarterly*, no. 78 (July 2015): 61–69.

Groen, Michael. *With the 1st Marine Division in Iraq, 2003: No Greater Friend, No Worse Enemy*. Quantico, VA: Marine Corps History Division, 2006.

Hall, Edward T. *The Silent Language*. Garden City, NY: Doubleday, 1959.

Hall, Edward T., and George L. Trager. *The Analysis of Culture*. Wash-

ington, DC: Foreign Service Institute and American Council of Learned Societies, 1953.
Harrell, Margaret C., Harry J. Thie, Peter Schirmer, and Kevin Brancato. *Aligning the Stars: Improvements to General and Flag Officer Management*. Santa Monica, CA: Rand, 2004.
Harris, Marvin. *Why Nothing Works: The Anthropology of Daily Life*. New York: Simon and Schuster, 1981.
Hays, MSgt Ronald E., USMC (Ret). *Combined Action: U.S. Marines Fighting A Different War, August 1965 to September 1970*. Quantico, VA: History Division, an imprint of Marine Corps University Press, 2019.
Henk, Dan. "An Unparalleled Opportunity: Linking Anthropology, Human Security, and the U.S. Military." Presentation at the Society for Applied Anthropology's 66th Annual Meeting, Vancouver, BC, 28 March–2 April 2006.
Hofstede, Geert. *Culture and Organizations: Software of the Mind*. New York: McGraw Hill, 1991.
———. *Culture's Consequences: International Differences in Work-Related Values*. Thousand Oaks, CA: Sage, 1980.
Hofstede, Geert, Gert Jan Hofstede, and Michael Minkov. *Culture and Organizations: Software of the Mind*, 3d ed. New York: McGraw Hill, 2010.
Hofstede, Geert, and Michael H. Bond. "Hofstede's Cultural Dimensions: An Independent Validation Using Rokeach Values Survey." *Journal of Cross-Cultural Psychology* 15, no. 4 (November 1984): 417–33.
Holmes-Eber, Paula. *Culture in Conflict: Irregular Warfare, Culture Policy, and the Marine Corps*. Stanford, CA: Stanford University Press, 2014.
Holmes-Eber, Paula, Erika Tarzi, and Basema Maki. "U.S. Marines' Attitudes Regarding Cross-Cultural Capabilities in Military Operations." *Armed Forces and Society* 42, no. 4 (2016): 741–51. https://doi.org/10.1177/0095327X15618654.
Holmes-Eber, Paula, and Maj Marcus J. Mainz. *Case Studies in Operational Culture*. Quantico, VA: Marine Corps University Press, 2014.
Human, Social, Culture, Behavior Modeling Program. *Human Social Culture Behavior Newsletter* 1, no. 1 (Spring 2009).

I Marine Expeditionary Force. "State of the Insurgency in al Anbar." Declassified, unpublished intelligence report. Ar-Ramadi, Iraq, 2006.

Ingold, Tim. "Introduction to Culture." In *Companion Encyclopedia of Anthropology*. Edited by Tim Ingold. New York: Routledge, 1994, 329–49.

Johnson, Jeannie L. "Fit for Future Conflict?: American Strategic Culture in the Context of Great Power Competition." *Journal of Advanced Military Studies* 11, no. 1 (Spring 2020): 185–208. https://doi.org/10.21140/mcuj.2020110109.

Katz, Eric. "Looking Back at the GSA Scandal: Did the Administration Overreact?" *Government Executive*, 26 January 2015.

Keiser, Lincoln. *Friend by Day, Enemy by Night: Organized Vengeance in a Kohistani Community*. Chicago, IL: Holt, Rinehart and Winston, 1991.

Kelly, George A. *A Theory of Personality: The Psychology of Personal Constructs*. New York: W. W. Norton, 1963.

Kipp, Jacob, Lester Grau, Karl Prinslow, and Capt Don Smith. "The Human Terrain System: A CORDS for the 21st Century." *Military Review* (September–October 2006): 8–15.

Kroenig, Matthew. "Why the U.S. Will Outcompete China." *Atlantic*, 3 April 2020.

Leeds-Hurwitz, Wendy. "Notes in the History of Intercultural Communication: The Foreign Service Institute and the Mandate for Intercultural Training." *Quarterly Journal of Speech* 76, no. 3 (1990): 262–81. https://doi.org/10.1080/00335639009383919.

Lewis, Adrian R. *The American Culture of War: The History of U.S. Military Force from World War II to Operation Iraqi Freedom*. New York: Routledge, 2007.

Liddell Hart, Basil H. *Thoughts on War*. London: Faber & Faber, 1944.

Lucas, George R. *Anthropologists in Arms: The Ethics of Military Anthropology*. Lanham, MD: Altamira Press, an imprint of Rowman & Littlefield, 2009.

Mackenzie, Lauren, Eric Gauldin, and Erika Tarzi. *Cross-Cultural Competence in the Department of Defense: An Annotated Bibliog-*

*raphy*. Quantico, VA: Center for Advanced Operational Culture Learning, 2018.

Mackenzie, Lauren, and John W. Miller. "Intercultural Training in the United States Military." In *The International Encyclopedia of Intercultural Communication*. Edited by Young Yun Kim and Kelly L. McKay-Semmler. Hoboken, NJ: John Wiley, 2017. https://doi.org/10.1002/9781118783665.ieicc0189.

Mackenzie, Lauren, and Megan Wallace. "Intentional Design: Using Iterative Modification to Enhance Online Learning for Professional Cohorts." In *Communicating User Experience: Applying Local Strategies Research to Digital Media Design*. Edited by Trudy Milburn. Lanham, MD: Rowman & Littlefield, 2015, 155–82.

McCarthy, Rory, and Ewan MacAskill. "US Steps Up Aggression in Tikrit." *Guardian*, 17 November 2003.

McCloskey, Michael J., Aniko Grandjean, Kyle J. Behymer, and Karol Ross. *Assessing the Development of Cross-Cultural Competence in Soldiers*. Arlington, VA: U.S. Army Research Institute for the Behavioral and Social Sciences, 2010.

McDonald, Daniel P., Gary McGuire, Joan Johnston, Allison Abbe, and Brian Selmeski. *Developing and Managing Cross-Cultural Competence within the Department of Defense: Recommendation's for Learning and Assessment*. Washington, DC: Defense Language Office, 2008.

McFate, Montgomery, and Janice H. Laurence. *Social Science Goes to War: The Human Terrain System in Iraq and Afghanistan*. London: Oxford University Press, 2015. https://doi.org/10.1093/acprof:oso/9780190216726.001.0001.

McLuhan, Marshall, and Quentin Fiore. *War and Peace in the Global Village*. New York: Bantam, 1968.

McMaster, H. R. "How China Sees the World." *Atlantic*, May 2020.

McNamara, Laura A., and Robert A. Rubinstein, eds. *Dangerous Liaisons: Anthropologists and the National Security State*, School for Advanced Research Advanced Seminar Series. Santa Fe, NM: School for Advanced Research Press, 2011.

McWilliams, CWO4 Timothy S., and LtCol Kurtis P. Wheeler, eds.

*The Anbar Awakening*, vol. 1, *American Perspectives*. Quantico, VA: Marine Corps University Press, 2009.

National Research Council. *Accelerating Technology Transition: Bridging the Valley of Death for Materials and Processes in Defense Systems*. Washington, DC: National Academies Press, 2004.

Patai, Raphael. *The Arab Mind*. New York: Scribner, 1973.

Pauly, John J. "A Beginner's Guide to Doing Qualitative Research in Mass Communication." *Journalism Monographs*, no. 125 (1991): 1–29.

Peacock, James, Robert Albro, Carolyn Fluehr-Lobban, Kerry Fosher, Laura McNamara, Monica Heller, George Marcus, David Price, and Alan Goodman. *Final Report, November 4, 2007*. Arlington, VA: AAA Commission on the Engagement of Anthropology with the US Security and Intelligence Communities, 2007.

Post, Kristin. "Interview with Jeffery Bearor." Translational Research Group, Center for Advanced Operational Culture Oral History Project, 6 March 2013.

Price, David H. *Anthropological Intelligence: The Deployment and Neglect of American Anthropology in the Second World War*. Durham, NC: Duke University Press, 2008.

———. "Cold War Anthropology: Collaborators and Victims of the National Security State." *Identities: Global Studies in Culture and Power* 4, nos. 3–4 (1998): 389–430. https://doi.org/10.1080/1070289X.1998.9962596.

———. *Weaponizing Anthropology: Social Science in the Service of the Militarized State*. Oakland, CA: AK Press, 2011.

Putnam, Linda L., and Samantha Rae Powers. "Developing Negotiation Competencies." In *Communication Competence*, Handbooks of Communication Science No. 22. Edited by Annegret F. Hannawa and Brian H. Spitzberg. Boston, MA: DeGruyter Mouton, 2015, 367–95. https://doi.org/10.1515/9783110317459-016.

Rasmussen, Louise J., and Winston R. Sieck. "Culture-General Competence: Evidence from a Cognitive Field Study of Professionals Who Work in Many Cultures." *International Journal of Intercultural Relations*, no. 48 (September 2015):75–90. http://dx.doi.org/10.1016/j.ijintrel.2015.03.014.

Rentsch, Joan R., Allison Gunderson, Gerald F. Goodwin, and Allison Abbe. *Conceptualizing Multicultural Perspective Taking Skills*. Arlington, VA: U.S. Army Research Institute for the Behavioral and Social Sciences, 2007.

Rentsch, Joan R., Ioana Mot, and Allison Abbe. *Identifying the Core Content and Structure of a Schema for Cultural Understanding*. Arlington, VA: U.S. Army Research Institute for the Behavioral and Social Sciences, 2009.

Ricks, Thomas E. " 'It Looked Weird and Felt Wrong'." *Washington Post*, 24 July 2006.

Robben, Antonius C.G.M. "Chaos, Mimesis and Dehumanization in Iraq: American Counterinsurgency in the Global War on Terror." *Social Anthropology* 18, no. 2 (2010): 138–54. https://doi.org/10.1111/j.1469-8676.2010.00102.x.

Rogers, Everett M. "The Extensions of Men: The Correspondence of Marshall McLuhan and Edward T. Hall." *Mass Communication and Society* 3, no. 1 (2000): 117–35. https://doi.org/10.1207/S15327825MCS0301_06.

Rogers, Everett M., William B. Hart, and Yoshitaka Miike. "Edward T. Hall and the History of Intercultural Communication: The United States and Japan." *Keio Communication Review*, no. 24 (2002): 3–26.

Rubinstein, Robert A. "Master Narratives, Retrospective Attribution, and Ritual Pollution in Anthropology's Engagements with the Military." In *Practicing Military Anthropology: Beyond Expectations and Traditional Boundaries*. Edited by Robert A. Rubinstein, Kerry B. Fosher, and Clementine K. Fujimura. Sterling, VA: Kumarian Press, 2012, 119–33.

Rush, Laurie, ed. *Archaeology, Cultural Property, and the Military*, Heritage Matters Series vol. 3. Woodbridge, UK: Boydell Press, 2010.

Sahlins, Marshall. "Two or Three Things that I Know about Culture." *Journal of the Royal Anthropological Institute* 5, no. 3 (September 1999): 399–422.

Salmoni, Barak A., and Paula Holmes-Eber. *Operational Culture for the Warfighter: Principles and Applications*. Quantico, VA: Marine Corps University Press, 2008.

Sargent, John F., Jr. *Defense Science and Technology Funding*. Washington, DC: Congressional Research Service, 2018.

Scales, MajGen Robert. "Army Transformation: Implications for the Future." Testimony to the House Armed Services Committee. Washington, DC, 21 July 2004.

Schirm, Allen L., Krisztina Marton, and Jeanne C. Rivard, eds. *Evaluation of the Minerva Research Initiative*. Washington, DC: National Academies Press, 2020.

Schmorrow, Capt Dylan. "Sociocultural Behavior Analysis and Modeling: Technologies for a Phase 0 World." Brief from the Office of the Assistant Secretary of Defense (Research and Engineering), 6 March 2013.

Schwille, Michael, Anthony Atler, Jonathan Welch, Christopher Paul, and Richard C. Baffa. *Improving Intelligence Support for Operations in the Information Environment*. Santa Monica, CA: Rand, 2020. https://doi.org/10.7249/RB10134.

Seligman, Martin. "The Hoffman Report, the Central Intelligence Agency, and the Defense of the Nation: A Personal View." *Health Psychology Open* 5, no. 2 (July–December 2018): 1–9. https://doi.org/10.1177/2055102918796192.

Selmeski, Brian R. "Bridging the Gap: Synchronizing Material and Behavioral Culture Programs." Presentation at the Annual Meeting of the Archaeological Institute of America, San Antonio, TX, 6–9 January 2011.

———. "Indigenous Integration into the Bolivian and Ecuadorean Armed Forces." In *Cultural Diversity in the Armed Forces*. Edited by Joseph Soeters and Jan Van der Meulen. London: Routledge, 2006, 48–63.

———. "Managing a Mixed Blessing: Achieving Educational Success while Serving Many Masters." Presentation at the Annual Meeting of the Southern Association of Colleges and Schools Commission on Colleges, Atlanta, GA, 3–6 December 2016.

———. *Military Cross-Cultural Competence: Core Concepts and Individual Development*, Armed Forces, and Society Occasional Paper Series No. 1. Kingston, ON: Royal Military College of Canada Centre for Security, 2007.

———. "Navigating the Slippery Slope: Balancing the Practical Benefits, Ethical Challenges, and Moral Imperative of Security Anthropology." Unpublished manuscript, 2007.

Simons, Anna. *21st-century Challenges of Command: A View from the Field*. Carlisle Barracks, PA: Strategic Studies Institute, 2017.

———. *21st Century Cultures of War: Advantage Them*. Philadelphia, PA: Foreign Policy Research Institute, 2013.

———. *The Company They Keep: Life Inside the U.S. Army Special Forces*. New York: Avon Books, 1997.

———. "Cynicism: A Brief Look at a Troubling Topic." *Small Wars Journal*, 16 February 2021.

———. *Networks of Dissolution: Somalia Undone*. New York: Routledge, 2018.

———. "On 'Military Anthropology'." *Parameters* 50, no. 3 (2020): 121–24.

Simons, Anna, Joe McGraw, and Duane Lauchengco. *The Sovereignty Solution: A Commonsense Approach to Global Security*. Annapolis, MD: Naval Institute Press, 2011.

Sims, Christopher J. *The Human Terrain System: Operationally Relevant Social Science Research in Iraq and Afghanistan*. Carlisle Barracks, PA: Army War College Press, 2015.

*Small Wars Manual*, NAVMC 2890. Washington, DC: Government Printing Office, 1940.

Soldz, Stephen, Bradley Olson, and Jean Maria Arrigo. "Interrogating the Ethics of Operational Psychology." *Journal of Community & Applied Social Psychology* 27, no. 4 (2017): 273–86. https://doi.org/10.1002/casp.2321.

Southern Association of Colleges and Schools, Commission on Colleges (SACSCOC). *Report of the Reaffirmation Committee: Air University Quality Enhancement Plan*. Maxwell Air Force Base, AL: Air University, 2009.

———. *Resource Manual for the Principles of Accreditation: Foundations for Quality Enhancement*. Decatur, GA: SACSCOC, 2018.

Sprague, Jo. "Why Teaching Works: The Transformative Power of Pedagogical Communication." *Communication Education* 42, no. 4 (1993): 349–66. https://doi.org/10.1080/03634529309378951.

Steen, Susan, Lauren Mackenzie, and Barton Buechner. "Incorporating Cosmopolitan Communication into Diverse Teaching and Training Contexts: Considerations from Our Work with Military Students and Veterans." In *Handbook of Communication Training: A Best Practices Framework for Assessing and Developing Competence*. Edited by J. D. Wallace and Dennis Becker. New York: Routledge, 2018, 401–13.

Sunseri, Amy. "Culture Summit Brings Nations Together, Promotes Understanding." U.S. Army, 9 March 2011.

Sycara, Katia, Geoff Gordon, Scott Atran, Jeremy Ginges, Michael Lewis, Cathy Tinsley, and David Traum. *Modeling Cultural Factors in Collaboration and Negotiation*. Research Triangle Park, NC: Army Research Office, 2014.

Tarzi, Erika. *Educating Marines: Reorienting Professional Military Education on the Target*. Quantico, VA: Translational Research Group Report, Center for Advanced Operational Culture Learning, Marine Corps University, 2018.

Tarzi, Erika, and Kerry B. Fosher. *Regional, Culture, and Language Familiarization Program Messaging*. Quantico, VA: Translational Research Group Report, Center for Advanced Operational Culture Learning, Marine Corps University, 2017.

Turney-High, H. H. *Primitive War: Its Practices and Concepts*. Columbia: University of South Carolina Press, 1991; original printing 1949.

Turnley, Jessica Glicken, and Aaron Perls. *What Is a Computational Social Model Anyway?: A Discussion of Definitions, a Consideration of Challenges, and an Explication of Process*. Albuquerque, NM: Advanced Systems and Concepts Office, Defense Threat Reduction Agency, 2008.

Tyson, Ann Scott. "Iraq's Restive 'Sunni Triangle'." *Christian Science Monitor*, 24 September 2003.

U.S. Air Force. *Culture, Region, and Language Flight Plan*. Washington, DC: U.S. Air Force, 2009.

———. "U.S. Air Force Fact Sheet: Cross-Cultural Competence." Air University, November 2017.

U.S. Department of Defense (DOD), *Defense Language Transformation Roadmap*. Washington, DC: DOD, 2005.

———. *Department of Defense Instruction 5160.70, Management of the*

    *Defense Language, Regional Expertise, and Culture (LREC) Program*. Office of the Under Secretary of Defense for Personnel and Readiness, 30 December 2016.

U.S. Navy. *Education for Seapower Strategy.* Washington, DC: Secretary of the Navy, 2020.

Vanden Brook, Tom. "$725m Program Army 'Killed' Found Alive, Growing." *USA Today*, 9 March 2016.

Vergun, David. "Great Power Competition Can Involve Conflict Below Threshold of War." Department of Defense, 2 October 2020.

———. *Renewed Great Power Competition: Implications for Defense—Issues for Congress*. Washington, DC: Congressional Research Service, 2021.

Wemer, David A. "U.S. Joint Chiefs Chairman Makes the Case for Keeping U.S. Troops in Europe." *New Atlanticist* (blog), Atlantic Council, 21 March 2019.

Whitlock, Craig. "U.S. Troops Tried to Burn 500 Korans in Blunder, Investigative Report Says." *Washington Post*, 27 August 2012.

Worth, Robert F. "Blast Destroys Shrine in Iraq, Setting Off Sectarian Fury." *New York Times*, 22 February 2006.

# INDEX

1st Marine Division, 21, 25n6
11 September 2001 (9/11), vii, 19–20, 69, 87, 90, 100–1

Abbe, Allison, 8, 16, 41, 125–41, 203, 206
Afghanistan, vii–viii, xiii, 3, 5, 21, 27, 38–39, 63, 90, 96, 98, 107n5, 137, 151, 165, 166n4, 167, 173–74, 176–80, 190n29, 198
Air Force Culture and Language Center (AFCLC), 5, 8, 57–58, 62, 64, 67, 69, 105, 107, 110, 112, 143, 205
airman/airmen (U.S. Air Force), ix, 44, 73n3, 106, 108, 116, 118, 120n40, 121
Air University, 7, 15–16, 44, 57, 64, 105–6, 111–12, 121–22
Air War College, 57, 105, 114, 168

al-Anbar Province, Iraq, 28–31, 35–36, 39
al-Fallujah, Iraq, 174
American Anthropological Association (AAA), 108n8, 115, 143, 214
anthropology, 6n3, 7, 58, 66, 68–70, 77–79, 81, 85–86, 95–96, 101–8, 115, 122, 133
area studies, 111, 113, 116, 120–21. *See also* regional studies
archive(s), xv, 14, 70, 197n36, 200, 206, 207n3
*Army Culture and Foreign Language Strategy*, 127
ar-Ramadi, Iraq, 28–32, 36, 174
ar-Rashidiya, Iraq, 21–22
Army Research Institute (ARI), 5, 12n13, 125–26, 135
Army Research Office (ARO), 130

Army Training and Doctrine Command (TRADOC) Culture Center, 5, 39, 126–29, 138
assessment, 113, 118–19, 121, 128, 137, 151–54, 159, 160n10, 189n28, 191, 200, 217

Barrett, Barbara M., 62
Bayji, Iraq, 24–26
Bearor, Jeffery, 8, 16–17, 162–84, 197n36, 201, 203
budgets, 11–12, 59, 61–62, 98, 116, 134, 136, 147, 151, 186, 196. *See also* funding
Bureau of Indian Affairs/Education, 68–69
bureaucracy, 4, 39, 159, 147–53. *See also* Deitchman, Seymour J.
Bush, George W., 90

Carter, James "Jimmy, 79
Center for Advanced Operational Culture Learning (CAOCL), U.S. Marine Corps, 5, 8, 16–17, 33–34, 39, 145, 163–64, 172n15, 175–76, 183n23, 184–86, 189–90, 192n30, 193n31, 196n35, 197n36, 200–1
Center for Army Leadership (CAL), 126–27
Center for Civil-Military Relations (CCMR), 98
Center for Language, Regional Expertise, and Culture (CLREC), U.S. Navy, 5
Center for Naval Analyses (CNA), 168, 169n8

chief of staff of the Air Force (CSAF), 102, 105, 112, 114
Cold War, xiv, 95, 111
Combat Hunter, 174–76
Combined Joint Special Operations Task Force (CJSOTF), 89, 96
Commission on the Engagement of Anthropology with the U.S. Security and Intelligence Communities (CEAUSSIC), American Anthropological Association, 214
Community College of the Air Force (CCAF), 69, 114, 121
computational modeling, 131–32, 211
communication, viii, ix, 15, 43–45, 49, 53–55, 57–61, 74, 114, 121, 133, 143, 147, 206
Connable, Ben, 7, 14, 19–40
contractors, 6, 37, 48, 119, 119, 155, 164, 186, 199–200
counterinsurgency (COIN), 13, 28, 36, 63, 91, 98, 107, 124, 128, 139–41, 151, 156, 165
*Counterinsurgency*, Field Manual 3-24, 129
Cross-Cultural Competence (3C), xi, 8, 16, 59, 63, 104–24, 126, 128, 143, 150. *See also* intercultural competence
culture dimensions, 130
culture-general, 49n15, 52, 117–20, 126–29, 143, 149, 176–77, 188, 192
culture-specific, 10, 48, 49n15, 116, 118, 149, 153, 176n19, 192, 211–12

236 ⊙ INDEX

cultural intelligence, 35–36, 39, 63, 109
curriculum, 41–55, 77, 79, 113–14, 120–21, 153, 158, 169, 181, 189n28, 211

Dallas, George, 8, 17, 41, 162–64, 184–201, 203
Defense Language and National Security Education Office (DLNSEO), 194n32
Defense Language Institute (DLI), Foreign Language Center (DLIFLC), 20, 106, 198
Defense Language Proficiency Test (DLPT), 106
Deitchman, Seymour J., 11–12
demand signal, 4, 119
Devlin, Col Peter, 36
distance education and training, 53
diversity and inclusion, 63, 73, 78

Education Command (EDCOM), 193
enlisted education, 42n1, 44n5, 57, 62, 67, 69, 91, 116, 121, 124, 138n22, 169n10, 170n12
Evans-Pritchard, E. E., 94

faculty, 5, 8, 15, 44, 53, 57–58, 62–63, 67, 72–73, 75–76, 78–79, 86, 90–91, 94, 97, 105–6, 108, 111–23, 141, 153, 157, 205
faculty senate, 64
foreign area officers (FAOs), 20, 22, 29–30, 106, 119, 141, 149n7

foreign language education, 106–7, 111, 116–19, 124, 126–27, 136, 138n22
Foreign Service Institute, 43, 109
Fosher, Kerry, 3–18, 33, 39, 41, 105, 107, 108n8, 119, 124, 142–61, 164–84, 203–8
full spectrum operations, vii
funding, 4n1, 5, 38n26, 76, 106, 112, 114, 117, 128, 130–31, 133, 136, 145, 150

Gelfand, Michele, 131
Global War on Terrorism (GWOT), 87, 89, 108
great power competition, 13, 63, 124, 140, 151
Green Berets, 78, 82, 84

Hall, Edward T., 42–43, 48, 109, 118
Helmand Province, Afghanistan, 154, 174
Henk, Daniel, vii–xii, 104–10, 112, 114, 116–17, 121, 123–24, 143
Hofstede, Geert, 130
Holmes-Eber, Paula, 46–47, 188
Human Relations Area Files (HRAF), 108, 188
Human, Social, Cultural, and Behavioral Modeling Program (HSCB), 131–32
Human Terrain System (HTS), xv, 12n13, 15, 19, 36–40, 66, 68, 115, 122, 138
Human Terrain Teams (HTT), 138

information operations, 13, 151

Ingold, Tim, 9
Inter-University Seminar (IUS) on Armed Forces and Society (IUSAFS), ix, 7, 105
intercultural competence, 128. *See also* cross-cultural competence
Iraq/Iraqi, vii–viii, xiii, 3, 7, 14, 19–31, 33–39, 63, 67, 90, 96, 98, 105, 107, 126, 128, 137, 151, 165–68, 171, 173–80, 190, 203

Jumper, Gen John P., 102

Kenya, 81–82
Keiser, Lincoln, 94
Kelly, George A., 164
Kelly, BGen John F., 19–20, 23–27
Khachadoorian, Angelle, 8, 15, 66–76, 203
kill chain, 35–36, 39
kinship, viii, 51, 210
Koran, 51n17

Language Enabled Airman Program (LEAP), U.S. Air Force, 117
language training, 4–5, 12, 14, 20–22, 45, 57–58, 61, 64, 67, 73, 75, 82, 102, 105–7, 109–12, 116–19, 123–29, 133, 136, 138, 143, 149–50, 158, 166–67, 169, 172, 175–76, 181, 194n32, 196–98, 205–6, 212
Language Training Detachment (LTD), 117
Liddell Hart, Basil H., 113
Lindsay, Gen James J., 84

Lorenz, LtGen Stephen R., 112–13, 117
LREC Symposium, 64

Mackenzie, Lauren, 3–18, 41–55, 64, 107n7, 184–201, 203–8
Maki, Basma, 46–47
maps/mapping, 24, 148, 210, 212
Marine Corps Combat Development Command (MCCDC), 163, 165, 184
Marine Corps Intelligence Activity (MCIA), 5, 8, 12n13, 33, 36, 145
Marine Corps University (MCU), 44, 53, 64, 105, 163, 183, 193n31, 200, 205
Marines (U.S.), ix, 7, 14, 20–21, 23–31, 43–52, 99, 154, 166, 169, 171–72, 174–78, 182, 186–92, 194, 200
Marlowe, David, 84n1
Marshall, Andrew, 101
Mattis, Gen James N., 21, 28, 50–51, 165, 167, 170–72, 181–83, 191, 194, 200–1
Maxwell Air Force Base, AL, 5, 7, 44, 57
McFate, Montgomery, 37
McLuhan, Marshall, 60
Mead, Margaret, 109
metacognition, 53
metrics, 37, 78, 106, 110, 113, 118, 137, 151–53. *See also* assessment
Military Anthropology Network (Mil_Ant_Net), 105
military culture center (MCC), 66, 70, 72, 74–76

mindset, 20, 22, 28–29, 35, 37–40, 47–48, 57, 200
modeling, 12n13, 130–32, 210–11. *See also* computational modeling
moral injury, 58–59
multidisciplinary university research initiative (MURI), 130–31

Naficy, Siamak, 102
National Defense Education Act of 1958, 111n16
Naval Postgraduate School (NPS), 15, 77–103, 167, 205
Netherlands, The, 108–9

Office of Management and Budget, 129
Office of Net Assessment (ONA), 97, 101
Office of the Secretary of Defense (OSD), 97, 99, 168, 178
officers, 19–30, 34, 36–39, 42n1, 44n5, 54, 57, 62, 71–72, 80, 82–89, 91, 95, 97, 99, 102, 104, 106, 119–21, 126, 128, 141, 149n7, 165, 169–70, 182, 184
online education and training, 50, 52, 62, 116, 121, 124, 153, 198n38
Operation Enduring Freedom (OEF), 190, 192
Operation Iraqi Freedom (OIF), 190, 192

Patai, Raphael, 35
Point IV Training Program, 109
post-traumatic stress disorder, 58–59
predeployment training/workups, ix, 31, 34, 100, 116, 118, 126, 128, 165, 169, 171, 176, 178–79, 183, 210, 212
professional military education (PME), xi, 15, 42, 44, 46, 53, 55, 57, 59, 61, 63–64, 71, 74, 77, 99, 107n5, 116, 141–42, 170–71, 183, 203

Quality Enhancement Plan (QEP), 110–23

Rasmussen, Louise J., 59
Regional Culture and Language Familiarization Program, U.S. Marine Corps (RCLF), 172, 191–92
Regional Security Education Program (RSEP), 99–100
regional studies/regional knowledge approach, 4, 12, 142, 154, 175n18. *See also* area studies
requirements, 50, 89, 111–12, 117, 124, 135, 139, 150, 157, 163, 165n3, 181, 187, 198
resilience, 58–59, 156
Roche, James G., 101
Rotary International Ambassadorial Scholar, 56–57

sailor (U.S.), ix, 99, 108
Scales, Gen Robert H., 178
science and technology, 4, 16, 130–32, 134
SEALs, 86, 95. *See also* Special Operations Forces
security cooperation, 1, 38, 151, 189

INDEX ◉ 239

Security Force Assistance Brigades, 138
Selmeski, Brian R., 7–8, 10, 16, 41, 104–24, 143, 203
Sieck, Winston R., 59
Simons, Anna, 7, 15, 77–103, 203, 205
simulations, 4, 118, 148, 210–11
Singapore, 109
small wars, 13, 165, 166n4, 171–72, 177, 181
Small Wars Center, 184
*Small Wars Manual*, 165–67, 180
social science/scientists, xiv, 11–12, 27, 32, 69, 71, 74, 79, 99, 125–42, 159
Society for Applied Anthropology, 104, 115
soldier (U.S.), ix, 25–26, 28–30, 37, 44, 80, 85, 108, 126, 128, 131–32, 178
Somalia, vii, 82, 84n1, 93, 100
Southern Association of Colleges and Schools, Commission on Colleges (SACSCOC), 111
Southwestern Indian Polytechnic Institute (SIPI), 69–70
special operations, 5, 77–78, 82–89, 91, 95, 97, 100, 141, 168
Special Operations Command (SOCOM), 83, 168. *See also* special operations
Special Operations Forces (SOF), 84, 141. *See also* special operations
Special Operations/Low-Intensity Conflict (SO/LIC), 77–78, 85, 95, 103

Special Operations Task Force (SOTF), 89, 96
Speyer, Arthur, 32–33, 36, 38
Steen, Susan, 8, 15, 56–65, 203
strategic empathy, 57
strategic-level
 communication, 147
 learning, 107, 120
 planning, 127, 129, 141, 206
 thinking, 97, 132
subject matter expert (SME), 4, 131, 153–56, 163, 170, 179, 183, 197, 199, 210–11.
Sycara, Katia, 131
Syria, vii, 177

tactical level, 107, 120, 128, 132, 137, 194
Tarzi, Erika, 46–47
Teaching and Learning Center, 64
Tiger, Lionel, 101
Tikrit, Iraq, 20, 22–27
Training and Doctrine Command (TRADOC) Culture Center, U.S. Army, 5, 19, 39, 126
Training and Education Command (TECOM), 165, 167–68, 170, 175, 180, 183, 193
transformation/transforming, x, 48
tribes, tribal, 21–25, 28–31, 68, 86, 210
Turney-High, H. H., 93

United States Air Force, viii–ix, xi, 5, 7, 16, 42, 46, 62, 67–69, 71–73, 85–86, 101–2, 105, 107, 115, 118, 120,

124, 137, 139, 143, 171, 188
United States Air Force Academy, 8, 67, 69–70
United States Army, ix, xv, 12n13, 15–16, 24–30, 37, 39, 42, 66–67, 78, 82, 85–86, 89, 104, 115, 125, 127–30, 133, 135–36, 138–40, 168, 171, 194, 198n35
United States Navy, 5, 12n13, 85n2, 86, 89n4, 99, 137, 139
United States Marine Corps, viii, ix, 15–16, 19, 21, 35, 38, 42, 49n15, 85n2, 159, 163–65, 166n4, 167–68, 169n9–10, 170n12, 171–72, 174n16, 176–77, 185, 187, 189–94, 197n36, 198

video series, 49–52
Vietnam conflict/war, x, xiv, 11, 27, 78, 128, 139, 165, 167, 172, 177, 181, 195, 197n36, 206

wasta, 52
White, LtCol Duffy W., 23
wisdom, 30, 55, 113, 200
World War II, x, xiv, 11, 95, 130, 146n6, 172n14, 195, 206

Yemen, vii, 99

# ABOUT THE AUTHORS

DR. ALLISON ABBE is professor of organizational studies at the U.S. Army War College in Carlisle, Pennsylvania. She previously served in a variety of government and research organizations, including the Institute for Defense Analyses in Alexandria, Virginia; the Office of the Director of National Intelligence; the interagency High-Value Detainee Interrogation Group; and the U.S. Army Research Institute for Behavioral and Social Sciences then located at Vienna, Virginia. She conducts applied research on the development, training, and management of national security personnel, with a focus on leadership and intercultural skills. Prior to entering government service, Abbe was a visiting assistant professor at George Washington University. She holds a PhD in social and personality psychology from the University of California, Riverside.

JEFFERY BEAROR is assistant deputy commandant for Manpower and Reserve Affairs (M&RA) and assists the deputy commandant in the operations and management of the M&RA Department ensuring integration of all its diverse human resource functions toward accomplishment of Marine Corps missions. The M&RA Department develops policies and programs across a broad range of functions directly supporting active and reserve component Marines, their family members, and civilian Marines. Appointed to the

Senior Executive Service in 2008, Mr. Bearor served as the executive deputy of the U.S. Marine Corps Training and Education Command at Quantico, Virginia. In October 2013, he was appointed as senior director for security for the Department of the Navy (DON) supporting the deputy under secretary of the Navy (DUSN) who performs duties as the senior executive and senior agency official for security for the DON. In this position he acted as senior advisor to DUSN on a wide variety of security policy, oversight, and resource matters. Mr. Bearor retired from active duty in the U.S. Marine Corps after serving for more than 30 years and has more than 46 years of military, Joint, interagency and intergovernmental leadership and operational experience. He joined the Marine Corps in 1975 following graduation from the University of Texas at Austin and was an infantry officer. He commanded four rifle companies, an infantry battalion, and the recruit training regiment at Parris Island, South Carolina. Mr. Bearor was a Marine Corps exchange officer with the British Royal Marine Commandos, graduated from their famed Commando Course, and was a military diver and parachutist. He also served as a detailee in the CIA's Counterterrorism Center, as the fleet Marine officer/force protection officer for the U.S. Fifth Fleet in Manama, Bahrain, and as the chief of current operations, J3, at the U.S. Central Command. He is a graduate of the University of Texas at Austin, the Marine Corps Command and Staff College, and the U.S. Army War College.

DR. BEN CONNABLE is a senior political scientist at Rand, where he leads research on warfare, regional issues, and human behavioral modeling, gaming, and simulation. He is also the director of research at DT Institute, a nonprofit peace and development organization in Washington, DC. Connable served as an infantry Marine, an intelligence officer, and a foreign area officer from 1988 to 2009. He deployed to Iraq three times in support of Operation Iraqi Freedom. He helped to stand up the Marine Corps' Center for Advanced Operational Culture Learning and co-led the Marine Corps' cultural intelligence program. He also represented the Marine Corps on a Department of Defense panel tasked with investing hundreds of millions of dollars to improve military cultural knowledge. He received his PhD in war studies from King's College London in 2016.

GEORGE DALLAS was commissioned in May 1978 after graduating from Shippensburg University in Pennsylvania with a bachelor of science degree in business administration. He has a master of arts degree (with distinction) in national security and strategic studies from the U.S. Naval War College in Newport, Rhode Island. Colonel Dallas was an artillery officer with numerous command and leadership positions within the operating forces and supporting establishment. Throughout his career, Colonel Dallas attended the Basic Officers Leadership and Advanced Field Artillery Schools at Fort Sill, Oklahoma; the Naval War College; and the U.S. Army War College in Carlisle, Pennsylvania. Colonel Dallas retired from active duty in September 2008 and assumed his position as director, Center for Advanced Operational Culture Learning, which he led until its closure in 2020.

DR. KERRY B. FOSHER is a sociocultural anthropologist whose work focuses on security organizations and their ability to integrate scientific knowledge and expertise. Within the Department of Defense (DOD), she has been the director of the U.S. Air Force cross-cultural competence project, the command social scientist for Marine Corps Intelligence Activity on Marine Corps Base Quantico, and the director of research for the Marine Corps' Center for Advanced Operational Culture Learning. She currently is the director of research for Marine Corps University. Prior to working with the DOD, Dr. Fosher was on the research faculty of the Dartmouth (now Geisel) School of Medicine in Hanover, New Hampshire, and a fellow at the Belfer Center for Science and International Affairs at Harvard's John F. Kennedy School of Government in Cambridge, Massachusetts. Her research has been supported by the National Science Foundation, the Roscoe Martin Fund, and through numerous projects sponsored by the Marine Corps. She publishes and presents regularly on ethics and the integration of science in military organizations. Fosher earned her PhD in anthropology from the Maxwell School of Citizenship and Public Affairs at Syracuse University in New York State.

DR. ANGELLE KHACHADOORIAN is an associate professor of anthropology in the Air Force Culture and Language Center

(AFCLC) at Air University on Maxwell Air Force Base, Alabama. She has taught a range of college and graduate-level courses for more than 20 years. She has been at the AFCLC for nine years and is affiliated faculty with the Air War College. Previously, Dr. Khachadoorian taught for the Bureau of Indian Affairs and was a distinguished visiting professor at the United States Air Force Academy in Colorado. Her research interests include Air Force organizational culture; Native American, indigenous, and tribal cultures, including the ways in which tribal groups maintain cultural continuity when intersecting with nation-states; and community memory and identity in the face of stressors and pressure to change or assimilate. She holds a PhD in anthropology from the University of New Mexico in Albuquerque.

DR. LAUREN MACKENZIE is a professor of military cross-cultural competence at Marine Corps University in Quantico, Virginia. She also chairs the Marine Corps University faculty council and serves as an adjunct professor of military/emergency medicine at the Uniformed Services University of Health Sciences in Bethesda, Maryland. Dr. Mackenzie's research interests revolve around the art and science of difficult conversations, and she has written a range of articles and book chapters devoted to such areas as end-of-life communication, the communication of respect, relationship repair strategies, and, most recently, the role of failure in education. She earned her MA and PhD in communication from the University of Massachusetts in Amherst and has taught intercultural and interpersonal communication courses throughout the DOD during the past 12 years.

DR. BRIAN R. SELMESKI is the chief of faculty affairs and an associate professor at Air University. His administrative duties include overseeing faculty-related policies and processes at the United States' largest military educational institution. As a faculty member, he teaches about Latin America, working across cultures, cultural property protection, and U.S. military culture. Since 2019, he has also chaired Air University's Public K-12 Education Working Group. Previously, he worked at the Canadian Defence Academy and Royal Military College of Canada in Kingston, Ontario, as well as the Latin American Faculty of Social Sciences and Pontifical Catholic University of Ecuador in Quito. His field research focuses on the military's relationship with civil society in Latin America and has been generously funded by the Social Science Research Council, the Senator J. William Fulbright Scholar Program, and the government of Canada's Human Security Fund. In the 1990s, Dr. Selmeski served as a U.S. Army officer, stationed primarily in Central America. He received his PhD in anthropology from Syracuse University's Maxwell School of Citizenship and Public Affairs.

DR. ANNA SIMONS is a professor emerita of defense analysis at the Naval Postgraduate School (NPS) in Monterey, California. She joined the Special Operations/Low-Intensity Conflict curriculum at NPS from 1998 to 2019, after six years as an assistant and then associate professor of anthropology at University of California, Los Angeles. Dr. Simons is the author of several books and multiple monographs: *Networks of Dissolution: Somalia Undone* (1995); *The Company They Keep: Life Inside the U.S. Army Special Forces* (1997); coauthored *The Sovereignty Solution: A Commonsense Approach to Global Security* (2011); *21st Century Cultures of War: Advantage Them* (2013); *21st Century Challenges of Command: A View from the Field* (2017); and has recently completed a monograph about military advising. She holds a PhD in social anthropology from Harvard University and an AB from Harvard College.

DR. SUSAN STEEN is an assistant professor of cross-cultural communication in the Air Force Culture and Language Center (AFCLC) at Air University on Maxwell Air Force Base, Alabama, where she has served as a faculty senator and recently completed a year as faculty senate president. Her scholarship involves intercultural, interpersonal, and organizational communication. Most recently, her work has focused on the role of cosmopolitan communication, deriving from Coordinated Management of Meaning (CMM) Theory, in bridging conflict and promoting intercultural competence. She serves on the board of the CMM Institute, an international organization devoted to advancing communication theory and practice as a means of shaping better social worlds. Dr. Steen earned her PhD in communication from the University of Southern Mississippi in 2007 and held a variety of positions in the field of international education prior to joining the AFCLC in 2015.